Springer Aerospace Technology

The series explores the technology and the science related to the aircraft and spacecraft including concept, design, assembly, control and maintenance. The topics cover aircraft, missiles, space vehicles, aircraft engines and propulsion units. The volumes of the series present the fundamentals, the applications and the advances in all the fields related to aerospace engineering, including:

- structural analysis,
- aerodynamics,
- aeroelasticity,
- aeroacoustics,
- flight mechanics and dynamics
- orbital maneuvers,
- avionics,
- systems design,
- materials technology,
- launch technology,
- payload and satellite technology,
- space industry, medicine and biology.

The series' scope includes monographs, professional books, advanced textbooks, as well as selected contributions from specialized conferences and workshops.

The volumes of the series are single-blind peer-reviewed.

To submit a proposal or request further information, please contact: Mr. Pierpaolo Riva at pierpaolo.riva@springer.com (Europe and Americas) Mr. Mengchu Huang at mengchu.huang@springer.com (China)

The series is indexed in Scopus and Compendex

Sergey Viktorovich Dvornikov ·
Alexander Fedotovich Kryachko ·
Igor Anatolyevich Velmisov ·
Dmitry Alexandrovich Zatuchny

Radio-Electronic Equipment in Civil Aviation

Construction and Maintenance

Springer

Sergey Viktorovich Dvornikov
St. Petersburg State University
of Aerospace Instrumentation
St. Petersburg, Russia

Alexander Fedotovich Kryachko
St. Petersburg State University
of Aerospace Instrumentation
St. Petersburg, Russia

Igor Anatolyevich Velmisov
St. Petersburg State University
of Aerospace Instrumentation
Leningrad region, Russia

Dmitry Alexandrovich Zatuchny
Moscow State Technical University of Civil Aviation
Moscow, Russia

ISSN 1869-1730 ISSN 1869-1749 (electronic)
Springer Aerospace Technology
ISBN 978-981-19-6198-4 ISBN 978-981-19-6199-1 (eBook)
https://doi.org/10.1007/978-981-19-6199-1

This Springer imprint is published by the registered company Springer Nature Singapore Pte Ltd.
The registered company address is: 152 Beach Road, #21-01/04 Gateway East, Singapore 189721, Singapore

Introduction

Safe transportation of people and valuable cargo is the main task of civil aviation at the present time. The complexity of implementing this function in accordance with the established requirements increases sharply in conditions of constantly increasing air traffic intensity and reducing the time to make the right decision.

In these conditions, it is necessary to ensure the accurate and reliable operation of radio navigation, radar and radio communication equipment systems in civil aviation, as well as to ensure their timely modernization.

Radio receivers are one of the main components in the operation of this equipment. It should be noted, that there are currently quite a few such devices. At the same time, the implementation of some of them in civil aviation is limited or even impossible due to economic factors and the lack of an appropriate material base.

The purpose of this book is to familiarize students of higher educational technical institutions of aviation profile, as well as undergraduates and postgraduates with the main types of radio receivers, that are used in the aviation industry.

In the Chap. 1 of the book, the issues of the influence of the electromagnetic environment on the operation of radio receivers are considered. At the same time, much attention is paid to the requirements for the structure and elements of the main receiving channel of the radio receiver.

In the Chap. 2, radio receivers of amplitude-modulated signals are considered, which are widely used in telephone and telegraph radio communication systems, radio telemetry and radio control, primarily due to the simpler implementation of modulation and demodulation devices.

The Chap. 3 of the book is devoted to the problems, associated with the implementation of pulse signal radios, widely used in radar and radio navigation, as well as in radio communication systems. At the same time, much attention is paid to issues, related to the possible distortion of pulse signals.

The Chap. 4 is devoted to the functioning of radio receivers with angular modulation. At the same time, much attention is devoted to the description of the methods of noise immunity of these devices. The principles of obtaining a coherent reference vibration in the presence of a carrier in the received signal are described.

The Chap. 5 of the book deals with issues, related to the functioning of long-range radio receivers, designed to receive radio signals in the frequency range, exceeding their bandwidth. Such devices include receivers of communication systems, receivers of radio engineering systems, using multi-frequency signals, panoramic, measuring radio control systems, etc. The main attention in this chapter is paid to the methods of setting up such receivers and describing the features of their structural schemes.

In the Chap. 6, receivers of optical systems are considered. At the same time, much attention is paid to their characteristics, structural diagrams and description of noise sources in such devices. A comparative evaluation of optical receivers was carried out.

The Chap. 7 of the book examines the measurement of the main characteristics of radio receivers, including sensitivity and distortion of signals. At the same time, issues, related to automatic gain control, frequency control and automation of testing are considered.

The textbook is based on lectures, given by the authors at the St. Petersburg State University of Aerospace Instrumentation and at the Moscow State Technical University of Civil Aviation.

Contents

Abbreviations

AD	Arithmetic device
AE	Antenna equivalent
AFA	Audio frequency amplifier
AFC	Audio frequency control
AFCS	Audio frequency control system
AGC	Automatic gain control
AM	Amplitude modulation
APCS	Automatic phase control system
ASK	Amplitude shift keying
BC	Correction block
BPALNA	A broadband path, an amplitude limiter and a narrowband amplifier
CA	Control automat
CE	Control element
CM	Character multiplier
CuA	The current amplifier
DB	Dynamic bandwidth
DC amplifier	Direct current amplifier
DCA	DC amplifier
DD	A deciding device
DL	Delay line
DRMS	The double root-mean-square value of the resulting detuning of the receive
EC REE	Electromagnetic compatibility of radio electronic equipment
EFS	Electronic or electromechanical frequency search
EMF	Electro moving force
F	Filter
FD	Frequency detector
FF	Frequency feedback
FM	Frequency modulation
FMS	Frequency modulation system

FLL	Frequency locked loop
FPI	Photon (quantum) radiation detectors
FSSB	The filter of single sideband
FSK	Frequency Shift Keying
G	Generator
GHF	Generators of high frequency
GLF	Generators of low frequency
HAGC	High-speed automatic gain control
HF amplifier	High frequency amplifier
HFS	Heterodyne frequency synthesizer
HPF	High-pass filter
IAGC	Inertial automatic gain control
IF	Intermediate frequency
IFA	Intermediate frequency amplifier
INAGC	Instantaneous automatic gain control
IR	Infrared radiation
IRF	Intermediate radio frequency
LAD	A linear amplitude detector
LFA	Low frequency amplifier
LLS	Laser location systems
LPF	Low-pass filter
MD	A measuring device
MRC	Main receiving channel
MRP	Main receiving path
MU	The mechanics unit
NC	A number comparator
NF	Noise figure
OF	An optical filter
OH	An optical heterodyne
OPM	Optical mixer
PAM	Pulse-amplitude modulation
PCM	Pulse-code modulation
PCM-AM	Pulse code modulation—Amplitude modulation
PCM-FM	Pulse code modulation—Frequency modulation
PCM—PM	Pulse code modulation—phase modulation
PCM—FM—PM	Pulse code modulation—Frequency modulation—Phase modulation
PCM—PFM—PM	Pulse code modulation—Pulse frequency modulation—Phase modulation
PD	Phase detector
PLF	Phase-locked frequency system
PLL	Phase-locked loop
PMT	Photomultiplier tubes
PPM	Pulse-phase (time) modulation
PSF	Pilot-signal filter

PSS	Pulse synchronization scheme
PWM	Pulse-width modulation
RD	Recording device
RFA	Radio frequency amplifier
RPSK	Relative phase shift keying
RR	Radio receivers
RRD	Radio receiving device
RRR	Range radio receivers
RS	Receiver sensitivity
SBF	Sideband filter
SI	Standard impulses
SLC	A square-law characteristic
SLO	The second local oscillator
SM	A semitransparent mirror
SPFI	Signal of the standard pulse former
SRC	Side receiving channel
SSB	Single sideband
SSG	Standard signal generator
SSM	Side-side modulation
TC	Technical condition
TD	A threshold device
TPI	Thermal radiation detectors
TR	Technical requirements
TSF	Threshold-forming
TS	Technical specifications
UHF	Ultra high frequency
ULF	Ultra low frequency
VA	Video amplifiers
VM	Voltage meter
VT	The video path

Chapter 1
Influence of Electromagnetic Situation on the Principles of Construction of the Main Reception Tract

1.1 Electromagnetic Interference

In the radio frequency range, a large number of radio systems for various purposes operate simultaneously, the transmitting devices of which, along with various high-frequency installations (industrial, scientific, medical, household and other purposes), are sources of unintentional electromagnetic interference.

An undesirable effect of electromagnetic energy is called electromagnetic interference, which worsens or may worsen the quality indicators of the functioning of radio electronic means [1].

The rapid quantitative growth of various sources of radiation, the number of which approximately doubles every ten years, the tendency to an increase in the power emitted by them and the wide possibilities of their spatial placement (from space to ground-based) lead to a progressive deterioration of the electromagnetic environment in the world. The need to ensure high-quality operation of radio systems in these conditions has determined the emergence of a new independent direction in radio electronics.

It is the electromagnetic compatibility of radio electronic equipment (EC REE), which is understood as the ability of radio electronic equipment to simultaneously function in real operating conditions with the required quality when exposed to unintentional interference [1].

The most common radio interference, which one has to meet when solving issues of EC REE, is the radiation of transmitting devices. Particularly dangerous is interference from stations that operate in the frequency bands allocated simultaneously to other stations. To reduce such interference, state coordination of the use of radio frequencies is carried out. However, in this case, the main radiation of a radio transmitting device can interfere with other radio equipment due to the frequency instability of the transmitter, non-observance of the rules for its operation, as well as due to randomly occurring specific conditions for the propagation of radio waves, leading to the appearance of interference signals at unplanned points in space or territory.

S. V. Dvornikov et al., *Radio-Electronic Equipment in Civil Aviation*, Springer Aerospace Technology, https://doi.org/10.1007/978-981-19-6199-1_1

Spurious (unwanted) emissions from transmitting devices resulting from nonlinear processes in the transmitter path are very dangerous sources of interference. These include:

- radiation at harmonics, i.e., at frequencies that are multiples of the frequency of the fundamental radiation;
- radiation at subharmonics, i.e., at frequencies an integer number of times lower than the frequency of the fundamental radiation;
- spurious emissions arising as a result of self-excitation due to parasitic couplings in the transmitter stages;
- intermodulation radiation due to the interaction on nonlinear elements of high-frequency the path of the transmitting device of the generated oscillations and the external electromagnetic field of other transmitting devices.

Interference is characterized by probability density and energy spectral density. In the general case, the interference at the input of the receiver can be represented as a certain voltage, the instantaneous value of which at any moment of time randomly amounts to any value from a number of definite values [2].

Analytically, the interference can be characterized by the frequency of occurrence of its instantaneous values. Let's call the frequency of occurrence at the input of the receiver the value of the interference U_i equal to the ratio of the number of measurements at which the interference had voltage U_i to the total number of measurements [3]:

$$\nu(U_i) = \frac{m_i}{\sum_{i=1}^{N} m_i}$$

then the expression

$$p(U_i) = \lim_{N\to\infty} \nu(U_i) = \lim_{N\to\infty} \frac{m_i}{\sum_{i=1}^{N} m_i}$$

is called the probability that the interference at any moment of time will take a value equal to U_i. The concept of probability characterizes random processes, the exact meaning of which is not known in advance. Such a process is the change in time of the interference voltage at the input of the receiver, the specific form of which is not known in advance. A random process can be represented by a number of discrete or continuous random variables that are a function of some parameters. Time is one of these parameters.

Let the probability that the interference voltage u_M at the input of the receiver at some point in time will take on a value lying within the range $U \le u_M < U + \Delta U$ is equal to $p(U,\ U + \Delta U)$. Then the limit of the ratio at $\Delta U \to 0$

$$\frac{p(U,\ U + \Delta U)}{\Delta U}$$

is called the density of the distribution of the probability of interference, if this limit exists:

$$W(U) = \lim_{\Delta U \to 0} \frac{p(U,\ U + \Delta U)}{\Delta U}$$

Then you can write $W(U)dU = p(U, U + \mathrm{d}U)$. When known $W(U)$, it is possible to determine the probability that the interference voltage will lie in the range between U_1 and U_2:

$$p(U_1 \leq u_M < U_2) = \int_{U_1}^{U_2} W(U)\mathrm{d}U$$

Graphically, the probability $p(U_1 \leq u_M < U_2)$ is numerically equal to the area under the curve $W(U)$, the abscissa and the ordinates U_1 and U_2. In addition, the interference at the input of the receiver is characterized by a probability distribution function, which is understood as the probability $F(U)$ that the instantaneous values it receives are equal U_3 to less than U:

$$F(U) = p(-\infty \leq U_3 < U) = \int_{-\infty}^{U} W(x)\mathrm{d}x$$

The energy spectral density of the interference $G(\omega)$ characterizes the intensity of its components in different parts of the spectrum, but does not give an idea of their initial phases. Therefore $G(\omega)$, it is impossible to reproduce the course of the function $u_M(t)$, for this, you need to know the spectral function $g(\omega)$ at a given time interval [4]:

$$g(\omega) = \frac{1}{2\pi} \int_{-\infty}^{\infty} U(t) \exp[-i\omega t]\mathrm{d}t$$

The modulus $g(\omega)$ characterizes the intensity, and the argument characterizes the phase of the spectral component of the oscillation $U(t)$.

Radio interference complicates the operation of the receiving device and creates the prerequisites for distorting the information it receives. The degree of influence of interference is determined by the technical perfection of the receiving device and its susceptibility to interference.

The susceptibility of a radio receiving device is called its property to react to radio interference, acting through the antenna and in addition to it (including through the screen) through the supply, control and switching circuits [1].

The susceptibility of a radio receiving device is determined by a number of its technical characteristics, the main of which are discussed in the following subsections.

The choice of the appropriate circuitry and obtaining the admissible values of the blocking coefficients, cross-distortion and intermodulation in the first stage of the receiver does not guarantee generally favorable characteristics of the entire broadband path, since out-of-band interference is amplified in it along with the signal. Therefore, the gain of the first stages of the receiver is made as small as possible. They try to resolve the resulting contradiction with the requirements for minimizing the receiver noise figure with the minimum allowable gain of the stages of the broadband path [5].

For the same purpose, with a useful signal exceeding the receiver sensitivity, it is desirable to reduce the gain of the broadband path. However, the use of cascades, in which regulation is carried out by changing the mode of their operation and which are widely used in AGC systems, in a broadband path is undesirable, since they degrade the linearity of its transfer characteristic. Here, as a rule, the functions of amplification and regulation are separated. To reduce the gain of the path, for example, linear controlled step diode-resistive dividers are used. In this case, with a decrease in the signal level, the level of the corresponding interference decreases simultaneously, which ensures a decrease in the influence of nonlinear effects on the operation of the receiver.

1.2 Single Signal Selectivity of a Superheterodine Receiver

One of the main characteristics of a receiver that determines its susceptibility is the single-signal selectivity characteristic, an approximate view of which is shown in Fig. 1.1. This characteristic is understood as the dependence of the signal level at the input of the radio receiver on the frequency of this signal for a given signal-to-noise ratio or a given signal level at the output of the radio receiver [1]. The figure shows the main, adjacent and most dangerous side channels of reception.

The main channel of a radio receiver is called the frequency band intended for signal reception. The side channels of reception are frequency bands that are outside the main reception channel and in which the signal passes to the output of the radio receiver [1]. The main reception channel with an average frequency f_0 is characterized by the bandwidth Δf_B and attenuation of adjacent σ_C reception channels f_{CH} and f_{CB}, the detuning Δf_C which are determined by service standards for frequency distribution. Side channels include an attenuated σ_m mirror channel, the center frequency of which is at the upper local oscillator setting $f_m = f_0 + 2f_{if}$, an intermediate frequency channel f_{if} and combination frequency channels, for example, $f_{K1,2} = 2f_G \pm f_{if}$, $f_{K3,4} = 0.5(f_G \pm f_{if})$. The attenuation of the adjacent channel is provided mainly by narrow-band filters of the intermediate frequency cascades, which form the characteristic of the frequency selectivity of the main channel.

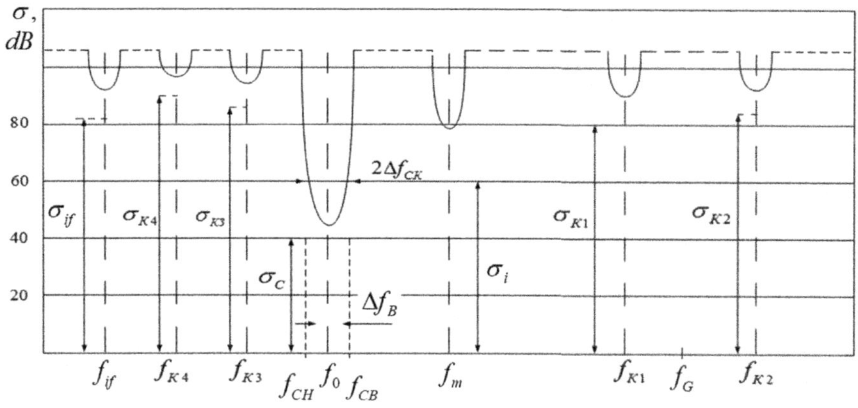

Fig. 1.1 An approximate view of the single-signal selectivity characteristic

Obviously, for a given squareness coefficient determined by the type and design of the bandpass filter, the attenuation of the adjacent channel will be the greater, the narrower the receiver bandwidth.

Ideally, the bandwidth of the receiver should be equal to the bandwidth Δf_{BS} of the received signal. However, the actual required bandwidth Δf_B is greater than the spectrum width due to possible random drifts of transmitter and local oscillator frequencies, as well as Doppler frequency shift. Indeed, if during the operation of the receiver the frequency of the local oscillator changes Δf_G by a certain amount, then the value of the combination frequency at the output of the converter will change by the same amount. Its new value will no longer correspond to the nominal value of the intermediate frequency Δf_{if} to which the IF amplifier path is tuned. The receiver in this case will be out of tune, and, as can be seen from Fig. 1.2, there will be frequency distortions of the signal. With a large drift of the local oscillator frequency, normal signal reception may be disrupted altogether. Similarly, transmitter carrier frequency instability and Doppler frequency shift appear. Practically

$$\Delta f_B = \Delta f_{BS} + \Delta f_{\text{drms}} + \Delta f_D,$$

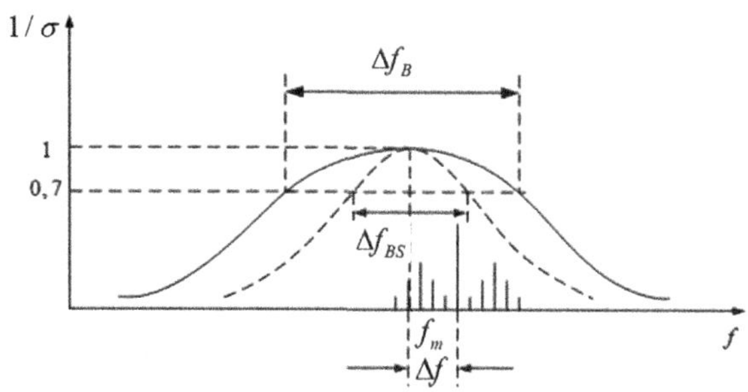

Fig. 1.2 Frequency distortion of the signal

where Δf_{drms} is the double root-mean-square value of the resulting detuning of the receiver r $\Delta f_{\text{drms}} = 2\sqrt{\alpha_C^2 f_C^2 + \alpha_G^2 f_G^2}$ (α_C, α_G)—the values of the relative instability of the frequencies of the transmitter and the local oscillator); Δf_D—Doppler frequency shift.

The need to obtain the required bandwidth imposes certain restrictions on the value of the intermediate frequency of the receiver. So, for example, in quartz or piezoceramic filters, the value of the maximum resonant frequency is limited by the design and minimum allowable dimensions of piezoresonators, and in filters with distributed selectivity, as shown in Chap. 5, by the minimum constructively feasible equivalent attenuation of their contours. As a result, the possible real values of the receiver intermediate frequency are bounded from above:

$$\Delta f_{if} \le f_{if\,\max} \tag{1.1}$$

The specific value $\Delta f_{if\,\max}$ is determined by the type of filters used in the IF path. Failure to satisfy inequality (1.1) leads to an expansion of the real bandwidth relative to the required one, and, consequently, to a decrease in the attenuation of the adjacent receive channel and the receiver sensitivity.

As can be seen from Fig. 1.3, for a given amplitude-frequency characteristic of the preselector, determined by the type and the number of filters tuned to the frequency of the signal, the attenuation of the image channel is the greater, the greater the value of the intermediate frequency of the receiver.

Difficulties in obtaining small squareness coefficients in input, especially tunable, filters, the need to increase their passbands compared to the required one to practically eliminate the effect of the preselector on the selectivity characteristic of the receiver as a whole, design constraints that do not allow obtaining relative filter passbands at meter and shorter wavelengths less than one to five percent of the value of the carrier frequency, limit from below the range of possible values of intermediate frequencies that meet the requirements of selectivity for the mirror channel [6]:

$$\Delta f_{if} \ge f_{if\,\min} \tag{1.2}$$

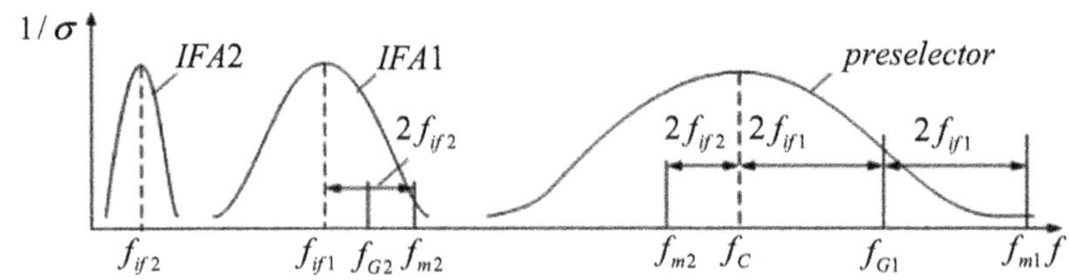

Fig. 1.3 The amplitude-frequency characteristic of the preselector

The specific value $f_{if\,\min}$ is determined by the parameters of the selectivity characteristics of the preselector. Failure to satisfy equality (1.2) will lead to deterioration of the selectivity for the mirror reception channel.

The boundary values of frequencies $f_{if\,\max}$ and $f_{if\,\min}$ are mutually independent, since they are determined by the design features of various bandpass filters of the receiving path. At $f_{if\,\max} \geq f_{if\,\min}$, intermediate frequencies f_{if} belonging to the set $(f_{if\,\min},\ f_{if\,\max})$ provide the required attenuation of both adjacent σ_C and mirror σ_m reception channels.

Inequality $f_{if\,\max} < f_{if\,\min}$ indicates the absence of a specific value f_{if} that simultaneously provides the specified attenuation σ_C and σ_m. The resolution of this contradiction is possible by applying double (or multiple) frequency conversion.

In a receiver with double conversion of frequency, as can be seen from Fig. 1.3, the received signal is sequentially converted in the first and second frequency converters:

$$f_{G1} - f_{=} f_{if1}, \qquad f_{G1} - f_{if1} = f_{if2}$$

With a large value of the first intermediate frequency in the preselector, it is possible to carry out the necessary attenuation of the first image reception channel, the frequency of which $-f_{m1} = f_C + 2f_{if1}$. At the second intermediate frequency (relatively low), the formation of the required resonance characteristics of the receiver and its bandwidth is provided, based on the necessary attenuation of the adjacent receiving channel. However, during the second frequency conversion, an additional mirror channel appears, located symmetrically with respect to the frequency of the second local oscillator $f'_{m1} = f_{if1} + 2f_{if2}$. Such a noise, passing through the preselector, will create a combination frequency at the output of the first converter

$$f_{G1} - f_{m2} = f_{G1} - (f_C - 2f_{if2}) = f_{if1} + 2f_{if2} = f'_{m2}$$

It is easy to see that the second image channel at the frequency can be attenuated in the required number of times with the appropriate choice of the type and parameters of the bandpass filters in the first intermediate frequency path.

1.3 Effects of Interference on Broadband Receiver Cascades

The characteristic of one-signal selectivity gives an idea of the attenuation of radio interference provided that the receiver operates in a linear mode, which corresponds to low levels of signal and interference at its input. The latter condition is practically not fulfilled in the current electromagnetic situation in military groupings, position areas of units and formations.

The real dynamic range of both signals and interference is very wide. In this case, the frequencies of radio interference may or may not coincide with the frequencies of the main and side channels of reception, and the levels of radio interference reach such values at which various nonlinear processes appear in the broadband path of the receiver tuned to the signal frequency, which significantly increases the receiver's susceptibility to radio interference. The analysis of nonlinear effects in a broadband path and the use of its results for the optimal synthesis of the circuits under consideration present significant difficulties, primarily due to the lack of effective universal research methods [7–9].

A significant simplification of the preliminary analysis can be achieved by dividing the conditionally broadband receiver paths into two types: paths with significant nonlinearity and paths with insignificant nonlinearity. Paths with significant nonlinearity are conventionally understood as such paths in which nonlinear transformations of the type of cutoff, limitation, key mode, etc. are performed. Conversely, paths in which such transformations do not occur are called paths with insignificant nonlinearity. One and the same path can operate in two modes of nonlinearity, depending on the level of signals at its input. So, in analog receiving and amplifying paths, operation with small input signals corresponds to a mode with insignificant nonlinearity, and for large ones—to a mode with significant nonlinearity.

Analysis of paths with significant nonlinearity, as a rule, is of a private, individual nature. General approximate solutions can be found for paths with insignificant nonlinearity.

Let us explain this on the model of a nonlinear four-port network (Fig. 1.4), which is equivalent to a single amplifier stage. In the general case, this four-port network corresponds to the system of nonlinear differential equations

$$\left.\begin{aligned} i_2 &= L_1[U_1, U_2], \\ i_2 &= L_2[U_1, U_2], \end{aligned}\right\} \tag{1.3}$$

where L_1 and L_2 are nonlinear operators.

If we consider the input current to be very small and neglect the inertia of the active element and the load response, then system (1.3) is transformed into one equation

$$i_2 = \varphi(U_1). \tag{1.4}$$

In this case, the transfer characteristic of the stage, which connects the input action and the response, is described by a simplified static characteristic of the amplifying device, which for the case of insignificant nonlinearity can be represented as a limited

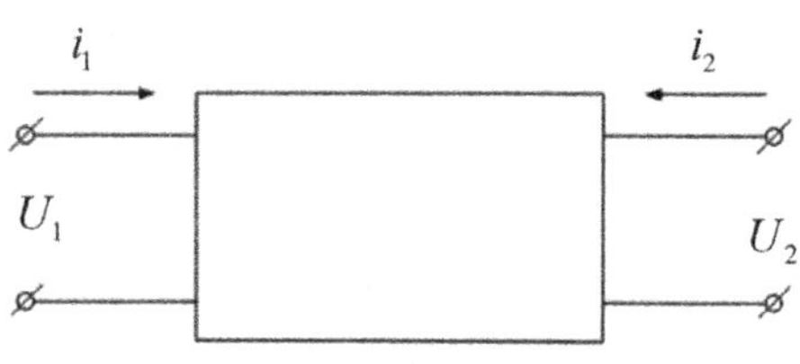

Fig. 1.4 The model of a nonlinear four-port network

power series with a degree of at least three. As shown by theoretical studies and experiments, this is the minimum number of members of the series, taking into account which allows you to get a complete picture of all the main nonlinear effects that arise in broadband receiving-amplifying paths [10–13].

Note that the analysis of nonlinear inertial circuits with the above assumptions does not allow obtaining accurate quantitative estimates of nonlinear phenomena in complex multielement paths, since it does not take into account the numerous external feedbacks widely used in modern microcircuitry, and the effects of interaction of nonlinear products taking into account the oscillation phases in multistage circuits. A more accurate and detailed presentation of these issues can be found in [2, 14, 15]. The current–voltage characteristic of the active element, shown in Fig. 1.5, is nonlinear, but in the operating mode in a fairly wide range from $+\Delta U$ to $-\Delta U$ around the bias point U_0, it is continuous and smooth, which is true for modern transistors and vacuum tubes. Assuming that the voltage varies from $+\Delta U$ to $-\Delta U$ around the original bias $U_1 = U_0$, we represent the output current of the active element in the form of a Taylor series:

$$i_2 = \varphi(U_0 \pm \Delta U) = \varphi(U_0) + \frac{\varphi'(U_0)}{1!}\Delta U + \frac{\varphi''(U_0)}{2!}\Delta U^2 + \frac{\varphi'''(U_0)}{3!}\Delta U^3 + \cdots$$

Considering that $\varphi(U_0) = i_0$ is the constant component of the output current, and $\varphi'(U_0) = S$, $\varphi''(U_0) = S'$, $\varphi'''(U_0) = S''$ is, respectively, the slope of the characteristic of the active element at the operating point, its first and second derivatives, we have

$$i_2 = i_0 + S\Delta U + \frac{1}{2!}S'\Delta U^2 \frac{1}{3!}S''\Delta U^3 + \cdots . \tag{1.5}$$

Suppose that the input of the amplifier under consideration is simultaneously influenced by the signal and noise voltages falling into the bandwidth of the broadband path:

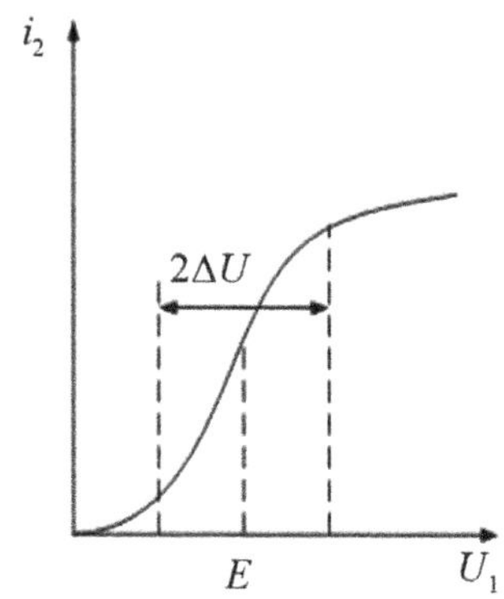

Fig. 1.5 The current–voltage characteristic of the active element

$$\Delta U = U_{TC} \cos \omega_C t + U_{TG} \cos \omega_G t. \tag{1.6}$$

Analysis of relation (1.5) after substituting into it the value ΔU from equality (1.6) shows that in this case the output current of the stage, due to the nonlinearity of its active element, contains combination components with frequencies $\pm p\omega_C \pm q\omega_G$. Of all the combinational frequency currents, only the components of the signal frequency to which the bandpass filter of the amplifier is tuned are of interest for further analysis. The currents of the remaining combination frequencies will be filtered out by the subsequent narrow-band path of the receiver.

The amplitude I_{TC} of the output current component of the stage i_2 has the form

$$I_{TC} = SU + \frac{S''}{8}U^3 + \frac{S''}{4}UU_G^2 + \ldots . \tag{1.7}$$

Relation (1.7) is an analytical expression for the vibrational characteristics of the cascade. In general, it is non-linear, since the amplitude of the output current is not proportional to the amplitude of the input signal. An approximate form of the vibrational characteristic is shown in Fig. 1.6. The nature of the graph is determined by the slope value *S*, its second derivative *S*, and the noise amplitude U_{TG}. In practice, for real amplifying devices, the second derivative of the slope is negative, which leads to a decrease in the signal amplitude with an increase in the magnitude of the noise.

Formally, this is explained by the fact that, as can be seen from Fig. 1.5, when exposed to interference, the value I_{TC} is determined not by the differential value of the slope S at the operating point, but by its integral, averaged value over the interval $U_0 \pm U_{TG}$, which (due to the upper and lower bends of the characteristic $i_2 = \varphi(U_1)$ will decrease relatively steep *S*.

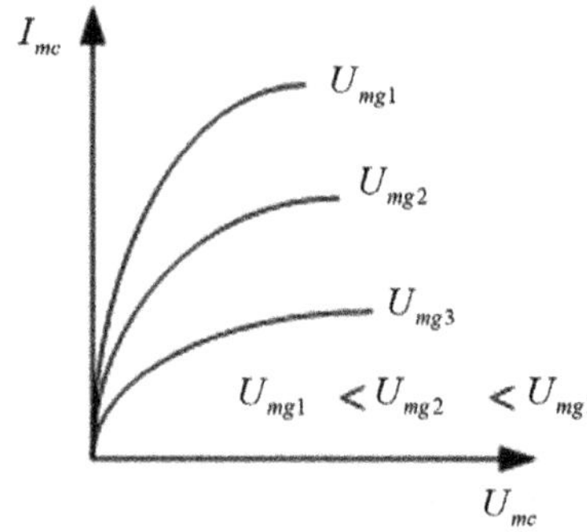

Fig. 1.6 An approximate form of the vibrational characteristic

1.3.1 Blocking

An analysis of the oscillatory characteristics, the graphs of which are shown in Fig. 1.6, shows that the amplitude of the signal at the output of the amplifier decreases the more, the greater the amplitude of the noise at its input.

Changing the signal level or signal-to-noise ratio at the output of the radio receiver under the action of radio interference, the frequency of which does not coincide with the frequencies of the main and side channels of the radio receiver, is called blocking [1]. For a quantitative assessment of the effect under consideration, a blocking coefficient-k_{bc} is introduced, which is understood as the ratio of the difference in signal levels at the output of the radio receiver in the absence and in the presence of radio interference at the input to the level of this signal in the absence of radio interference [16–18]:

$$k_{bc} = \frac{U_{\text{out}} - U_{\text{out}\,bc}}{U_{\text{out}}} = \frac{I_{TC} - I_{TC\,bc}}{I_T C}, \tag{1.8}$$

where U_{out}, I_{MC} are the values of the output signal in the absence of interference; $U_{\text{out}\ bc}$, $I_{TC\,bc}$—the same, in the presence of interference.

Obviously, in the absence of blocking $k_{bc} = 0$, and with complete blocking, when the signal level in the presence of interference is zero, $k_{bc} = 0$.

With a strong excess of the noise level over the signal (which is of the greatest practical interest), its oscillatory characteristics (1.7) can be written in the form

$$I_{TC} \approx SU + 0.25S''UU_G^2. \tag{1.9}$$

Hence, in accordance with formula (1.8),

$$k_{bc} \approx \frac{1}{4}\frac{S''}{S}U_{TG}^2. \tag{1.10}$$

Obviously, the greater the level of radio interference or the closer its frequency is to the tuning frequency of the radio receiver, the stronger the blocking effect is [19].

This phenomenon is quantitatively assessed by the characteristic of the frequency selectivity of the radio receiver for blocking (Fig. 1.7), which means the graph of the dependence of the amount of radio interference at the input of the receiver on its frequency for a given blocking coefficient and the presence of a useful signal with a level equal to the sensitivity of the radio receiver.

Often, for a numerical assessment of the frequency selectivity of a receiver for blocking, the concept of a dynamic range of a radio receiver for blocking is introduced D_{bc}. It is defined as the ratio of the level of radio interference U_G to the level of the useful signal $U_C = E_{AO}$ corresponding to the sensitivity of the radio receiver at a given blocking factor and frequency detuning of radio interference with respect to

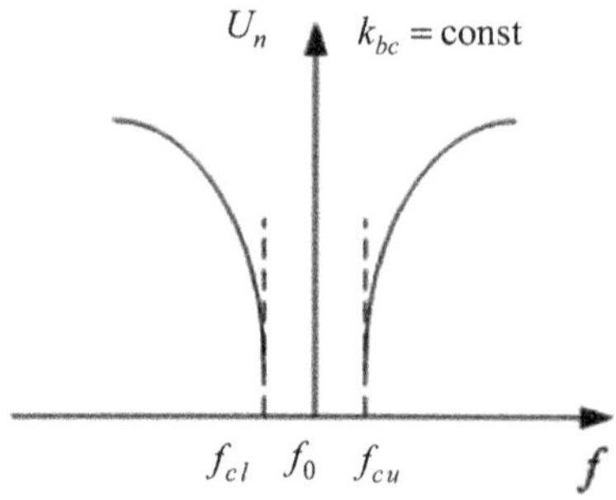

Fig. 1.7 The characteristic of the frequency selectivity of the radio receiver for blocking

the average frequency Δf of the main channel of the radio receiver. This ratio is expressed in decibels:

$$D_{bc} = 20\lg\left(\frac{U_G}{E_{AO}}\right). \quad (1.11)$$

In most cases, the value is taken to be equal to the detuning to the adjacent channel. Depending on the type of radio receiver or its purpose, the dynamic range for blocking in the adjacent channel (at $\Delta f = \Delta f_C$) is set by state standards or departmental norms.

Physically, the degradation of reception quality when blocking occurs mainly for three reasons. First, due to a decrease in the gain in the broadband path of the receiver, the signal at the output decreases, which is unacceptable at the input signal level corresponding to the sensitivity, since in this case the decrease in gain is not compensated by the operation of the Automatic gain control (AGC) system. Secondly, as a result of a decrease in the gain of the first stages, the noise figure of the radio receiver increases, which leads to a deterioration in the signal-to-noise ratio at its output. Thirdly, the value of the signal-to-noise ratio changes due to the appearance of noise modulation, the essence of which is that the noise signal of the frequency f_G plays the role of a local oscillator in a nonlinear system of a broadband path, which is transformed, as it were, into a frequency converter.

In this case, all the components of the white noise at the input of the receiver, the frequencies of which f_N satisfy the inequalities

$$f - 0.5\Delta f_G \leq n f_G \pm f_N \leq f_C + 0.5\Delta f_G,$$

after transformation with appropriate weighting coefficients fall into the main receiver channel, degrading the signal-to-noise ratio at its output.

1.3.2 Cross Modulation

Let us assume that the wanted signal and interference acting at the input of the broadband channel of the receiver are tone-modulated:

$$U_{TC}(t) = U(1 + m_C \cos \Omega_C t), U_{TG}(t) = U_{TG}(1 + m_G \cos \Omega_G t).$$

where m_C and m_n is the modulation depth of the signal and interference, respectively; Ω_C and Ω_G is the modulation frequency of the signal and interference.

In accordance with expression (1.9), the amplitude of the signal frequency current at the output of the broadband path changes in time according to the following law:

$$\begin{aligned} I_{TC}(t) &= SU_{TC}(1 + m_C \cos \Omega_C t) \\ &\quad + 0.25S''U_{TC}(1 + m_C \cos \Omega_C t)U_{TC}^2(1 + m_G \cos \Omega_G t)^2 \end{aligned} \tag{1.12}$$

After elementary transformations, setting $m_G^2 \ll I$ and $m_G m_C \ll I$, we have

$$\begin{aligned} I_{TC}(t) &\approx SU_{TC}(1 + m_C \cos \Omega_C t) \\ &\quad + 0.25S''U_{TC}U_{TG}^2(1 + m_C \cos \Omega_C t + 1 + m_G \cos \Omega_C t) \end{aligned}$$

Comparing relation (1.12) with expression (1.9), we see that with modulated interference, in addition to the blocking effect determined by the term $0.25S''U_{TC}U_{TG}^2$, a new effect appears—signal modulation by interference (the last term). The process of transferring interference modulation onto the signal carrier is called cross-distortion [14].

Cross distortion is understood as a change in the structure of the signal spectrum at the output of the radio receiver under the action of the signal and modulated radio interference, the frequency of which does not coincide with the frequencies of the main and side channels of the radio receiver [1].

The mechanism for transferring modulation from an interference carrier to a signal carrier is shown in Fig. 1.8. If the signal and noise are not modulated, then the amplitude of the signal current at the output I_{mC} will also have modulation, but its value will decrease due to blocking.

When modulating the interference, its amplitude relative to the carrier will vary from $+m_G U_{TG}$ to $-m_G U_{TG}$ with the modulation frequency Ω_G. As a result, the current amplitude of the signal frequency $I_{TC}(t)$ not at the output of the broadband path, and, consequently, the output voltage, will also change with the frequency of the noise envelope change.

Quantitatively, cross-distortion is estimated by the cross-distortion coefficient k_{cd}, which is understood as the ratio of the level of spectral components $U_{\Omega G}$ resulting from cross-distortion to the signal level at the receiver $U_{\Omega C}$ output for given radio interference and signal parameters [1].

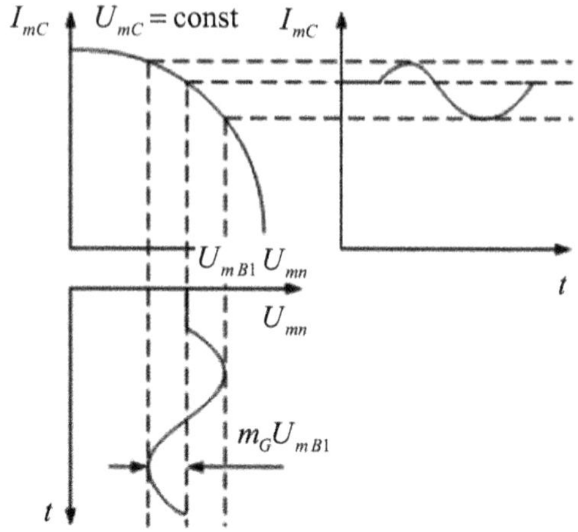

Fig. 1.8 The mechanism for transferring modulation from an interference carrier to a signal carrier

Assuming that the output voltage of the receiver is proportional to the change in the amplitude of the corresponding currents, choosing from expression (1.12) the frequency components Ω_Φ and Ω_C, we have

$$I_{T\Omega_C} = Sm_C U + 0.25S'' U_G^2 m_C U \approx Sm_C U,$$

$$I_{T\Omega_G} = 0.5S'' U m_G U_G^2,$$

where $I_{T\Omega_C}$, $I_{T\Omega_G}$ is the amplitude of the output low-frequency component, respectively, of the signal current and the interference frequency.

From here

$$k_{cd} = \frac{I_{T\Omega_G}}{I_{T\Omega_C}} \approx \frac{1}{2}\frac{S''}{S}\frac{m_G}{m_C} U_{TG}^2. \tag{1.13}$$

The amount of cross-distortion can also be estimated by the characteristic of the frequency selectivity of the radio receiver for cross-distortion, which is understood as the graph of the dependence of the level of modulated radio interference on frequency at a given cross-distortion factor and a signal corresponding to the sensitivity [1].

In some cases, to assess cross-distortion, the concept of the dynamic range of a radio receiver for cross-distortion is used, which is defined as the ratio (in decibels) of the level of modulated radio interference at the input of the radio receiver to the level of the useful signal, equal to the sensitivity of the receiver, at given cross-distortion factors and frequency detuning modulated radio interference in relation to the average frequency of the main receiving channel.

Since the mechanism of interference during blocking and cross-modulation is the same, in technical conditions the parameters are usually normalized either by blocking or by cross-distortion.

1.3.3 Intermodulation

In today's complex electromagnetic environment, it is very likely that two or more interference signals will arrive at the input of the broadband path of the radio receiver at the same time. In this case, the nonlinearity of the broadband path leads, as follows from the theory of frequency conversion, to the appearance of a number of combination frequencies at its output. Some of them may be equal to or close to the receiver tuning frequency. The resulting interference signal, falling into the main receiving channel, is amplified and appears at its output.

The occurrence of interference at the output of a radio receiver, when two or more radio interference signals are exposed to its input, the frequencies of which do not coincide with the frequencies of the main and side receiving channels of the radio receiver, is called intermodulation [1].

Intermodulation products, when exposed to the input of the receiver of two interference signals, are dangerous if the interference frequencies satisfy the inequalities

$$f_C 0.5\Delta f_G \leq |\pm n f_{G1} \pm m f_{G2}| \leq f_C + 0.5\Delta f_0. \quad (1.14)$$

The sum $n + m$ of the numbers determines the order of the intermodulation products.

Consider a possible way to assess the impact of intermodulation interference for the simplest case, corresponding to the second order combination, for which $n = m = I$. In this case, at the input of the broadband path, which is equivalent to a four-port network with insignificant nonlinearity, there are two noise signals applied relative to a certain operating point determined by the bias voltage U_0:

$$u = U_0 U_{TG1} \cos \omega_{G1} t = U_{TG2} \cos \omega_{G2} t$$

As is known from the theory of frequency conversion, the value of the combinational component of the current at the output of the nonlinear element is proportional to the conversion slope, which, with a simple conversion, is numerically equal to half the amplitude of the first harmonic of the slope change curve of the nonlinear element, which occurs under the action of the local oscillator voltage. In this case, one of the interference signals, for example, the first, can play the role of a local oscillator and the second—the role of a signal. Assuming, in the first approximation (Fig. 1.9), the change in the slope of the nonlinear four-port network to be linear in the interval, we have

$$I_{m\,\text{int}} \approx S_H U_{mG2} = 0.5 S' U_{mG1} U_{mG2}. \quad (1.15)$$

For a numerical assessment of the phenomenon under consideration, the concept of the intermodulation coefficient in the receiver is introduced, which means the ratio of the level of interference resulting from intermodulation in the radio receiver

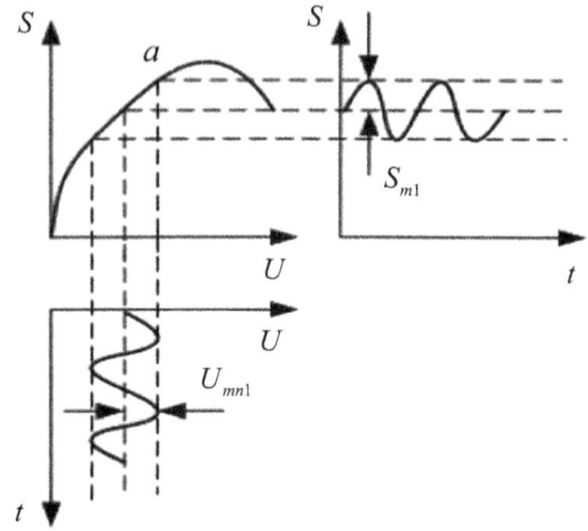

Fig. 1.9 The change in the slope of the nonlinear four-port network

to the signal level corresponding to the sensitivity of the radio receiver, while the interference and the signal are determined at the output of the radio receiver [1]:

$$k_{int} = \frac{U_{\text{out int}}}{U_{\text{out}}} = \frac{I_{m\ \text{int}}}{I_{mC}}, \tag{1.16}$$

where $U_{\text{out int}}$, $I_{m\ \text{int}}$ is the level of radio interference caused by intermodulation U_{out}, I_{mC} are the normal signal levels at the output of the radio receiver.

In the case considered above

$$k_{\text{int}} = \frac{I_{m\ \text{int}}}{I_{mC}} = \frac{S_H U_{mG2}}{SU_{mS}} = \frac{1}{2}\frac{S'}{S}\frac{U_{mG1}U_{mG2}}{U_{mC}}$$

In expression (1.16), the value corresponds to the input signal equivalent to the sensitivity.

The characteristic of the frequency selectivity of a radio receiver for intermodulation is the dependence of the magnitude of equal intensity signals at the input of the radio receiver on the frequency of one of them at a given intermodulation coefficient in the radio receiver [1].

Intermodulation interference is more pronounced the closer the interference signal frequencies are to the tuning frequency of the receiver and the lower the order of the intermodulation products. The first condition provides a small attenuation of interference in the broadband path of the receiver and the second, with equal amplitudes of interference, increases the conversion "efficiency" [15].

To assess the danger of intermodulation distortion, it is necessary to determine the possible frequencies of interference signals and establish a realistic relationship between the signal and interference levels. Based on the relation (1.14), taking into account that the receiver bandwidth, as a rule, is much less than the signal frequency ($\Delta f_B \ll f_C$), we have

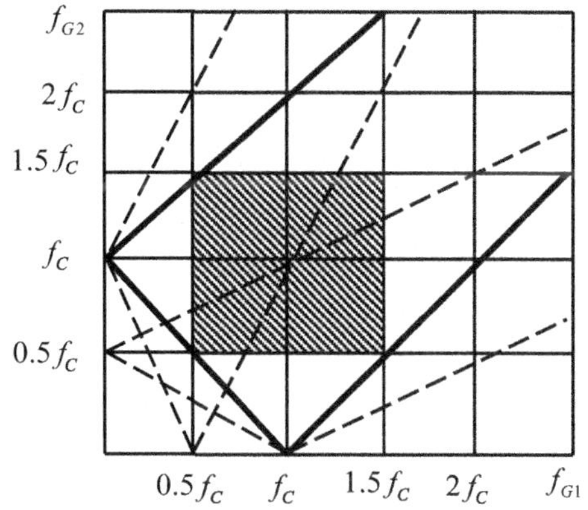

Fig. 1.10 A diagram, built according to the formula (1.17)

$$|\pm n f_{G1} \pm m f_{G2}| \approx f_C. \tag{1.17}$$

Figure 1.10 shows a diagram built according to the formula (1.17). In this diagram, the frequency of the receiver tuning lies at the intersection of the abscissa and ordinate, each of which is equal f_C. A frequency range is built around the tuning frequency, the boundaries of which correspond to the relative bandwidth of the broadband high-frequency channel of the receiver, equal to 1.0. In practice, it can be assumed that the interference whose frequencies go beyond this area will be greatly attenuated by the preselector and will not create intense intermodulation interference.

Based on the analysis of the diagram, it can be assumed that the most dangerous will be intermodulation interference of the third order of the form $2f_{G1} - f_{G2}$ or $2f_{G2} - f_{G1}$.

In this regard, in the technical specifications for radio receiving devices, as a rule, intermodulation distortions are normalized due to combinations of the third order, which are often called combinations of the "two-one" type.

1.4 Requirements to the Structure and Elements of the Main Receiving Channel of the Radio Receiver

As shown in the previous subsections, the receptivity and sensitivity of the receiving device is significantly dependent on its selectivity. Therefore, one of the most important problems solved by the rational construction of the linear path of the receiver (the main receiving channel—MRC) is to select a signal while simultaneously affecting the receiver input as a useful signal and sufficiently strong in-band and out-of-band electromagnetic interference [16].

The attenuation of interference penetrating the output of the receiver through adjacent receiving channels is ensured by a rational choice of bandpass filters in the intermediate frequency path based on the required bandwidth and the squareness coefficient of the resonance curve. The design features of these bandpass filters, with stringent requirements for their selectivity characteristic, can limit the value of the intermediate frequency of the receiver from above, which, at high signal frequencies and real characteristics of filters tuned to the signal frequency, worsens the attenuation of the side channels and, in particular, the mirror receive channel. Elimination of this contradiction is provided by the use of multiple frequency conversion in the receiver.

In a complex electromagnetic environment, when evaluating the single-signal selectivity of a receiver, in a number of cases, it is impossible to use the "small signal" approximation. So, for example, a side receive channel with a central frequency can be very dangerous $f_G = f_C + 0.5 f_{if1}$, which, after the first frequency conversion according to the second harmonics of the local oscillator and interference, enters the main receive channel:

$$2f_{G1} - 2f_G = 2f_{G1} - 2\left(f_C + 0.5 f_{if1}\right) = 2(f_{G1} - f_C) - f_{if1} = f_{if1}.$$

The receiver's exposure to out-of-band interference is determined by the levels of blocking factors, crosstalk and intermodulation. Therefore, one of the important tasks in the construction of the MRC, in particular the broadband path of the received frequency, is the all-round weakening of nonlinear effects arising in this path with the simultaneous effect of a signal at the input of the receiver and sufficiently strong out-of-band interference.

From consideration of nonlinear phenomena in the stages of a broadband path, it follows that high multi-signal selectivity is provided in cases where the active elements of the amplifying stages have small ratios S''/S and S'/S. It should be noted that in modern electronic devices it is impossible to select a mode in which both of these ratios are minimized simultaneously. Therefore, the broadband path must contain elements that maintain a high linearity of the transfer characteristic in a large dynamic range of input signals and interference. For this, in modern transistor amplifiers of the microwave range, push–pull circuits on powerful microwave transistors with several linearizing linear negative feedback circuits are widely used. Similar circuits are used in the lower frequency ranges. For the same purpose, diode balanced, ring and double ring mixers are widely used as the first frequency converters, up to the hectometer wavelength range [2].

The choice of the appropriate circuitry and obtaining the admissible values of the blocking coefficients, cross-distortion and intermodulation in the first stage of the receiver does not guarantee generally favorable characteristics of the entire broadband path, since out-of-band interference is amplified in it along with the signal. Therefore, the gain of the first stages of the receiver is made as small as possible. They try to resolve the resulting contradiction with the requirements for minimizing the receiver noise figure with the minimum allowable gain of the stages of the broadband path [17].

For the same purpose, with a useful signal exceeding the receiver sensitivity, it is desirable to reduce the gain of the broadband path. However, the use of cascades, in which regulation is carried out by changing the mode of their operation and which are widely used in AGC systems, in a broadband path is undesirable, since they degrade the linearity of its transfer characteristic. Here, as a rule, the functions of amplification and regulation are separated. To reduce the gain of the path, for example, linear controlled step diode-resistive dividers are used. In this case, with a decrease in the signal level, the level of the corresponding interference decreases simultaneously, which ensures a decrease in the influence of nonlinear effects on the operation of the receiver.

References

1. Belousov AP (1959) Calculation of the noise figure of radio receivers. State. publishing house of defense. prom-sti, M., 135p
2. Bogdanovich BM (1980) Nonlinear distortions in receiving and amplifying devices. Svyaz, Moscow, 280p
3. Vorobiev VI (1983) Optical location for radio engineers. Radio and communication, M., 176p
4. Goldenberg LM, Matyushkin BD, Polyak MN (1985) Digital signal processing: a handbook. Radio and communication, M., 312p
5. GOST 27002 - 83. Reliability in technology. Terms and definitions
6. GOST 93611-79. Electromagnetic compatibility of radio electronic means. Terms and definitions
7. Gutkin LS, Chentsova OS (1958) Transient processes in the system: high-frequency amplifier – detector. Radio Eng 11:18–23
8. Ditkin VA, Prudnikov AP (1965) A guide to operational calculus. Higher. shk., M., 466p
9. Evtyanov SI (1948) Transient processes in receiving-amplifying circuits. Radio and communication, M., 210p
10. Zernov NV, Karpov VG (1972) Theory of radio engineering circuits. Energiya, Leningrad, 816p
11. Vasilchenko NV, Borisov VA, Kremenchug LS, Levin GE (1983) Measurement of parameters of optical radiation receivers (Kurbatova LN, Vasilchenko NV (eds)). Radio and communication, M., 320p
12. Martynov VA, Selikhov YuI (1980) Pan-frame receivers and spectrum analyzers (Zavarin GD (ed)). Sov. radio, M., 352p
13. Pervachev SV (1982) Radioautomatics: textbook for universities. Radio and communication, M., 296p
14. Pestryakov VB, Kuzenkov VD (1985) Radio engineering systems. Radio and communication, M., 376p
15. Bankov VN, Barulin LG, Zhodzishsky MI et al (1984) Radio receivers. (Barulina LG (ed)). Radio and communication, M., 272p
16. Korostelev AA, Klyuev NF, Miller YuA et al (1978) Theoretical foundations of radar: textbook for universities (Dulevich VE (ed)). Soviet radio, M., 608p
17. Xyuz RS (1980) Logarithmic amplifiers. Sov.radio, Moscow, 352p
18. Zhodzishsky MI, Sila-Novitsky SYu, Prasolov VA et al (1980) Digital phase synchronization systems (Zhodzishsky MI (ed)). Soviet radio, M., 208p
19. Vimberg GP, Vinogradov YuV, Fomin AF et al (1972) Energy characteristics of space radio lines (Zenkiewicz OA (ed)). Sov. radio, M., 436p

Chapter 2
Radio Receiving Devices of Amplitude-Modulated Signals

2.1 Brief Description of Signals. Features of Radio Receivers Amplitude-Modulated Signals

Signals with amplitude modulation (AM) and amplitude shift keying (ASK), despite their relatively low noise immunity compared to signals with angle modulation, are widely used in telephone and telegraph radio communication systems, radio telemetry and radio control, due primarily to simpler implementation of modulation and demodulation devices. The simplest amplitude-modulated signal can be written as a ratio [1]

$$u_{AM} = U_{mC}[1 + m\cos(\Omega \mathrm{t} + \varphi)]\cos\omega_C \mathrm{t}, \tag{2.1}$$

where U_{mC} is the amplitude of the carrier vibration; Ω – modulation frequency; m—modulation factor; φ—the initial phase of the modulation frequency; ω_C—carrier frequency.

When a carrier is modulated with one frequency (harmonic) Ω, the signal spectrum contains three spectral components, which can be obtained by expanding expression (2.1):

$$\begin{aligned} u_{AM} = U_{mC}\cos\omega_C t + \frac{1}{2}mU_{mC}\cos[(\omega_H - \Omega)t - \varphi] + \\ + \frac{1}{2}mU_{mC}\cos[(\omega_C + \Omega)t + \varphi]. \end{aligned} \tag{2.2}$$

The spectrum width of the simplest AM signal is equal to the doubled modulation frequency, i.e. $\Delta\omega_C = 2\Omega$.

In the case of modulation of a signal carrier with a complex wave, having modulation frequencies from $\Omega_{\min}$ to $\Omega_{\max}$, in the AM signal spectrum there are a carrier and two side bands containing the same information. In this case, the width of

S. V. Dvornikov et al., *Radio-Electronic Equipment in Civil Aviation*, Springer Aerospace Technology, https://doi.org/10.1007/978-981-19-6199-1_2

the AM signal spectrum is determined by the maximum modulation frequency, i.e. $\Delta\omega_C = 2\Omega_{max}$.

A mode of operation using amplitude modulation in radio communications is called a mode $A3$. One of the main disadvantages of this mode is that the carrier frequency with the highest radiated power does not carry any information. Therefore, in order to improve the energy performance of radio links with AM, one or another form of modified AM signal is used. These modified AM signals include a two-sideband signal with an attenuated or suppressed carrier (operating mode $A3B$), a single-sideband signal with a carrier (operating mode), a single-sideband signal with a weakened carrier (operating mode $A3H$), and a single-sideband signal with a fully suppressed carrier (mode work $A3J$). From the energy point of view and from the point of view of the occupied frequency band, the modes $A3A$ and $A3J$ are the most rational modes of operation. In this case, the signal spectrum width is equal $\Delta\omega_C = \Omega_{max}$ for the mode $A3A$ and $\Delta\omega = \Omega_{max} - \Omega_{min}$ for the mode. It should be noted that the use of a single-sideband signal increases the immunity of radio communications to signal fading in the ionosphere and troposphere [2].

In modern communication systems, discrete (usually binary) signals are widely used to transmit information. With the help of such signals, telegraph messages, data from computing centers, quantized continuous messages, etc. are transmitted.

In particular, amplitude keying is widely used to provide auditory reception of telegraph signals. In this case, two modes of operation are possible—$A1$ and $A2$. In the mode $A1$, which is called telegraphy with amplitude shift keying, symbol 1 of the original message corresponds to the transmission of a carrier wave during the duration of the message and symbol 0 corresponds to the absence of oscillations (pause).

In a mode $A2$ called ASK tone telegraphy, during the transmission of symbol 1, the carrier wave is additionally modulated with a tone frequency F_{TF} voltage of the order of 800–1200 Hz. The widest spectrum of a telegraph signal is inherent in a periodic sequence of elementary pulses and pauses. In this case, the frequency of F_M manipulation, numerically equal to the frequency of the first harmonic of the periodic sequence of messages, is determined by the relation $F_M = \frac{1}{2}\tau_E$, where τ_E is the duration of the elementary message.

The real width of the signal spectrum for the modes $A1$ and $A2$ is determined by the expressions, respectively

$$\Delta\omega_C = 2n\Omega_M, \cdots \Delta\omega_C = 2(F_M n + F_{TF}),$$

where n is the number of the manipulation frequency harmonic, which depends on the signal registration method [3, 4].

When registering by the integral method $n = 1$, when registering by the method of shortened contact $n = 3 \div 5$. It should be noted that the width of the signal spectrum in the modes $A1$ and $A2$ is always narrower than the signal spectrum width in the mode $A3$. This is the main advantage of the operating modes $A1$ and $A2$. Structural diagrams of receivers for receiving signals in modes $A2$ and $A3$ are practically the same.

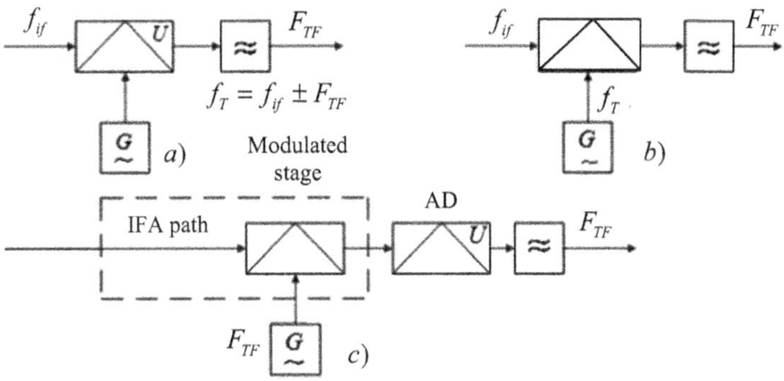

Fig. 2.1 **a** The method of heterodyne detection. **b** The additional conversion method. **c** The method of tone modulation

In the case of receiving signals in the mode $A1$, the structural diagram of the receiver will differ from the above-mentioned structural diagrams by the presence of additional elements that make it possible to receive the ASK signal by ear.

Reception of signals $A1$ by ear is possible by methods of heterodyne detection, additional frequency conversion and tone modulation in the receiver.

When receiving by the method of heterodyne detection (Fig. 2.1a), two oscillations are applied to a conventional amplitude detector: the voltage of a radio signal with amplitude shift keying from the common radio path and the voltage of a special telegraph local oscillator, the frequency of which differs from the intermediate one by a frequency f_{if} of $F_{TF} = 800 \div 1200$ Hz (i.e. $f_T = f_{if} \pm F_{TF}$).

Oscillations of two close frequencies f_{if} and f_T form beats, the envelope of which changes with the difference frequency F_{TF}. After the detector in the audio frequency amplifier (AFA), a filter with a narrow passband is placed, while the average frequency of the filter should correspond to the optimal tone frequency of 1000 Hz [5].

When receiving by the additional conversion method (Fig. 2.1b), a separate frequency converter is installed. The mixer input receives frequency f_{if} fluctuations from the common radio path and the frequency of the telegraph local oscillator $f_T = f_{if} \pm F_{TF}$, which differ by the frequency of the tone $F_{TF} = 800 \div 1200$ Hz. The voltage of the difference frequency after the converter is supplied to the ultrasonic amplifier for amplification and filtering using a band-pass filter with an average frequency of $F = 1000$ Hz [2, 6].

In the case of reception by the method of tone modulation (Fig. 2.1c), in one of the cascades of the AFA of the main receiving path (MRP), the received signal $A1$ is modulated with a tone of $F_{TF} = 800 \div 1200$ Hz from a special audio frequency generator. As a result, an oscillation of the form is obtained at the output of the modulated IF amplifier stage $A2$, which is detected in a conventional amplitude detector.

It should be noted that usually receivers provide the ability to receive various signals. Therefore, in them, the MRP of signals, in which the necessary amplification,

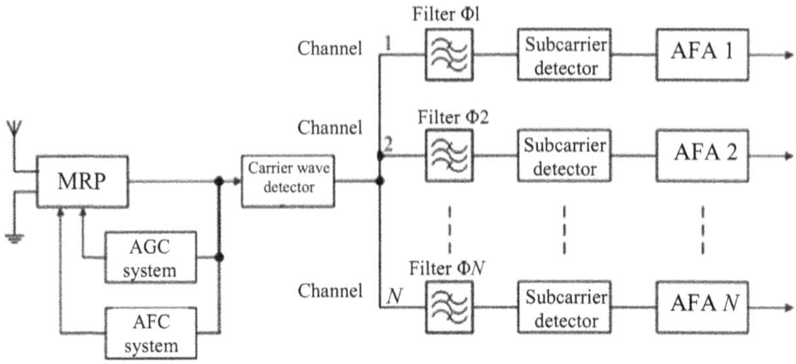

Fig. 2.2 The scheme of a multichannel receiver with frequency division multiplexing and modulation of the AM–AM type

filtering and conversion to the optimal intermediate frequency are provided, must be common. In the MRP, depending on the type of the received signal, the bandwidth adjustment can be provided.

To meet the high requirements for sensitivity and selectivity that are imposed on modern AM signal receivers, these receivers are built according to a superheterodyne circuit with single or multiple frequency conversion. Moreover, they can be single-channel or multichannel. In multichannel receivers, frequency or time division of the channels is used. As an example, consider the scheme of a multichannel receiver with frequency division multiplexing and modulation of the AM–AM type (Fig. 2.2).

In multichannel frequency division receivers, the signal is modulated twice: first at the subcarrier and then at the carrier. A doubly modulated signal is denoted, for example, as follows: AM–AM, FM–AM, SSM–AM, etc. In this case, the first letters characterize the modulation law of the subcarrier, and the second—the carrier. Amplitude modulation, frequency modulation, single sideband modulation, etc. can be used as the primary modulation, Fig. 2.3.

With modulation of the AM–AM type, a signal arrives at the input of the receiver, the spectrum of which is shown in Fig. 2.3a. This signal, after amplification and conversion in the MRP, is fed to the carrier amplitude detector, at the output of which a signal with a spectrum is shown in Fig. 2.3b. This baseband signal arrives at the channel filters $\omega_{\Omega 1}, \omega_{\Omega 2}, \ldots, \omega_{\Omega N}$, each of which is tuned to the frequency of its own subcarrier. The subcarrier selected in this way with its own spectrum is fed to the subcarrier detector, the type of which depends on the type of modulation. At the output of the detector, a primary message appears (Fig. 2.3c), which, after the corresponding amplification in the audio frequency amplifier, enters the terminal device. It should be noted that both in single-channel and multichannel receivers, to maintain a constant normal output voltage and for reliable reception of signals with random drifts of the transmitter and local oscillator frequencies, AGC and AFC systems are used [7].

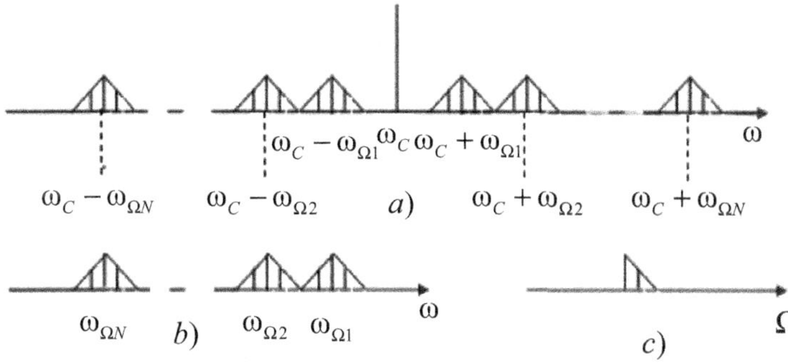

Fig. 2.3 Using amplitude modulation, frequency modulation and single sideband as the primary modulation: **a** the modulation of the AM–AM type; **b** the spectrum of the signal at the output of carrier amplitude detector; **c** a primary message

2.2 Distortions of Amplitude-Modulated Signals in the Main Reception Tract

In the main reception path, both nonlinear, and linear signal distortions are possible. Nonlinear distortions of the signal are caused by the nonlinearity of the characteristics of amplifying and converting devices used in this path. If the selector is not sufficiently selective, unwanted effects such as blocking, cross modulation and intermodulation may occur due to the nonlinearity of these characteristics.

The physical essence of these phenomena and their quantitative assessment will be considered in sufficient detail below. Therefore, in this case, we will not dwell on these phenomena in detail. We only note that in order to reduce nonlinear distortions, it is advisable to improve the selectivity of the preselector and to choose active elements of the MRP with good linearity of characteristics. The reason for the linear distortion of the signal is the unevenness of the amplitude-frequency characteristic of the amplifying path [8, 9]. To analyze these distortions in the main receiving path, we will assume that it receives a signal of the form described by relation (2.1). In this case, we will assume that the amplitude-frequency characteristic of the MRP is an even function and the phase-frequency characteristic is odd and linear within the bandwidth, i.e.

$$\left.\begin{aligned} K(\omega_0 + \Omega) &= K(\omega_0 - \Omega), \\ \varphi(\omega_0 + \Omega) &= -\varphi(\omega_0 - \Omega). \end{aligned}\right\} \tag{2.3}$$

It should be noted that a properly designed receiver with small detunings almost always satisfies condition (2.3). Let us first consider the case when the frequency of the carrier signal ω_C is equal to the average frequency ω_0 of the receiver's frequency response (Fig. 2.4).

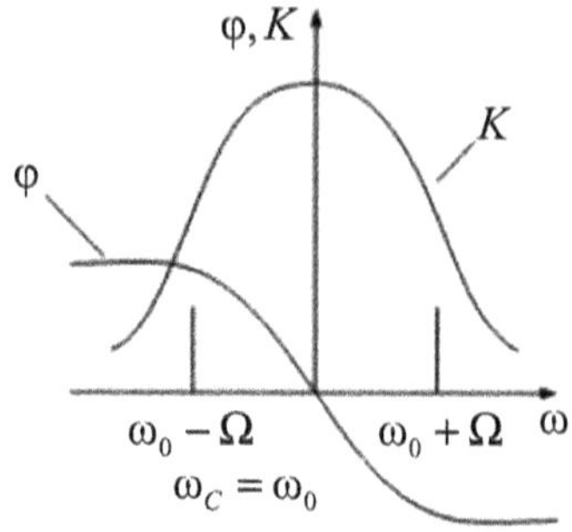

Fig. 2.4 The case when the frequency of the carrier signal is equal to the average frequency of the receiver's frequency response

The voltage at the output of the MRP is determined by the spectral method, according to which each spectral component of the AM signal represented by expression (2.2), when passing through the MRC, receives gain and phase shift, depending on the frequency of the corresponding spectral component. With this in mind, the voltage at the output of the MRC can be written in the form

$$U_{out} = U_{mC}K(\omega_0)\cos\omega_0 t + \frac{1}{2}mU_{mC}K(\omega_0+\Omega)\cos[(\omega_0+\Omega)t+\varphi$$
$$+\varphi(\omega_0+\Omega)] + \frac{1}{2}mU_{mC}K(\omega_0-\Omega)\cos[(\omega_0-\Omega)t-\varphi+\varphi(\omega_0-\Omega)]. \quad (2.4)$$

Taking into account condition (2.3) after transformations, expression (2.4) can be written as follows:

$$U_{out} = U_{mC}K(\omega_0)\left\{1+\frac{K(\omega_0+\Omega)}{K(\omega_0)}m\cos[\Omega t+\varphi+\varphi(\omega_0+\Omega)]\right\}\cos\omega_0 t. \quad (2.5)$$

Comparison of expressions (2.1) and (2.5) makes it possible to find out by distinguishing the voltage at the output of the MRC from the voltage at its input. Since the information at AM is contained in changes in amplitudes, it is of the greatest interest to compare the signal envelope at the input and output of the MRC [10].

From relation (2.5) it can be seen that the modulation coefficient of the output voltage

$$m_{out} = \frac{K(\omega_0+\Omega)}{K(\omega_0)}m. \quad (2.6)$$

Since, in accordance with Fig. 2.4, $K(\omega_0+\Omega) < K(\omega_0)$ then $m_{out} < m$ and, consequently, when the AM signal passes through the MRC, the modulation coefficient of the output voltage decreases.

Due to the linearity of the phase response, all spectral components of the input oscillation at the output of the MRC receive a phase shift proportional to their frequency. The consequence of this is that the envelope of the signal at the output of

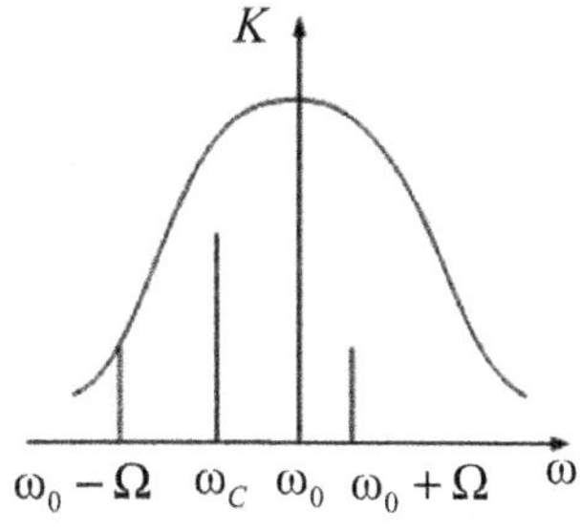

Fig. 2.5 Inaccurate tuning of the MRC to the AM signal

the MRC is delayed relative to the envelope of the input voltage by the delay time t_m, which is determined by the ratio of the phase shift of the upper lateral oscillation to the angular modulation frequency [11]: $t_m = \frac{\varphi(\Omega)}{\Omega}$.

In the case when the amplitude-frequency characteristic of the MRC is two-humped, i.e. when the inequality is satisfied $K(\omega_0 + \Omega) > K(\omega_0)$, overmodulation of the signal at the output of the MRC may occur, which will lead to the appearance of nonlinear distortions after detection.

With inaccurate tuning of the MRC to the AM signal carrier frequency (Fig. 2.5), the voltage spectrum at its output becomes asymmetric. One sideband is amplified more than the other, resulting in linear distortion.

In some cases, especially when the frequency response of the MRC is close to rectangular, complete suppression of one side component is possible. Then, in the spectrum of the output voltage, there will practically be two spectral components with frequencies ω_C and $\omega_C + \Omega$, between which beats occur. In this case, the envelope of the total voltage at the MRC output does not change according to a sinusoidal law, which leads to the appearance of nonlinear distortions after detection. In addition, it should be noted that in Fig. 2.5 in this case, the phase shifts that the lateral components with respect to the carrier oscillation receive when passing through the MRC turn out to be disproportionate to their frequency, and this leads to the appearance of nonlinear phase-frequency distortions [12].

Thus, in order to avoid distortions in the MRC, it is necessary to ensure, within its bandwidth, which exceeds the width of the spectrum of the received signal, the uniformity of the frequency response, the linearity of the phase response, and also ensure accurate tuning to the frequency of the received signal.

2.3 Detecting Two High-Frequency Vibrations with Different Carrier Frequencies

In practice, the detection of two high-frequency oscillations arises, for example, in cases when, as a result of insufficient selectivity of the MRP, a noise arrives at the detector, in addition to the useful signal, or when a high-frequency oscillation from a special local oscillator is additionally supplied to the detector, which occurs, as noted above, with heterodyne reception amplitude-keyed telegraph signals by ear [13].

Let us consider the features of the detector operation under these conditions.

Suppose that two unmodulated signals with frequencies ω_1 and ω_2 are supplied to the detector:

$$\left.\begin{aligned} u_1 &= U_{m1}\cos\omega t, \\ u_2 &= U_{m2}\cos\omega t. \end{aligned}\right\} \tag{2.7}$$

The voltage at the output of the detector is determined by the amplitude of the summed input voltage. Therefore, to assess the result of the impact of signals determined by the formula (2.7) on the detector, we consider the law of change in the amplitude of the total oscillation. It is known that the sum of two sinusoidal voltages with different frequencies forms a voltage, the amplitude and phase of which change with a beat frequency Ω_b equal to the difference in the frequencies $(\omega_2 - \omega_1)$ of the summed oscillations [4, 5]. The vector diagram of the beats and the timing diagram for their envelope are shown in Fig. 2.6. Here, the oscillation with greater amplitude U_{m1} and frequency ω_1 is represented as a vector $0O_1$. The second vibration with lower amplitude U_{m2} and frequency ω_2 is represented as a vector $0A$. $0A$.

The amplitude $U_{m\Sigma}$ of the total oscillation is determined by the geometric sum of the vectors U_{m1} and U_{m2}. The end of the vector $U_{m\Sigma}$ is located on the circle that the vector $0A_1$ describes as it rotates around the point 0.

Since the arc of the internal combustion engine is greater than the arc of the EDC (Fig. 2.6a), the amplitude of the total voltage in the time interval is greater than the amplitude of the voltage U_{m1}, and on the time t_1 interval, on the contrary $U_{m\Sigma}<U_{m1}$, (Fig. 2.6b), i.e. the envelope of the total voltage turns out to be asymmetric. In this case, the asymmetry of the envelope will decrease with a decrease in the voltage amplitude U_{m2} in comparison with U_{m1} and $U_{m2} \ll U_{m1}$, when the envelope of the resulting voltage is very close to a sinusoid [14, 15].

Consider the result of amplitude detection $U_{m\Sigma}$ for two cases. In the first case, we will say that the detector is inertial with respect to the beat frequency Ω_b, i.e.

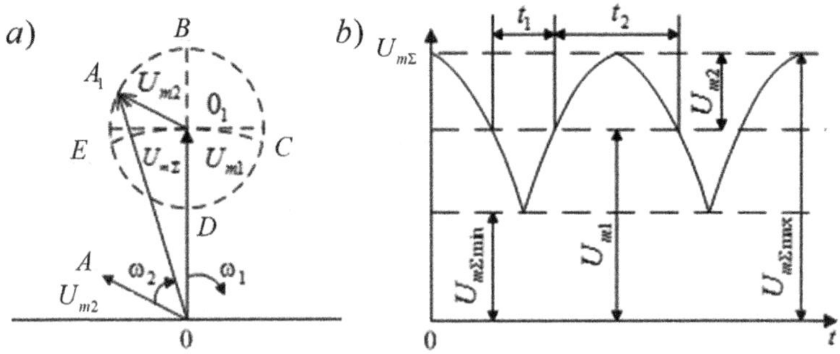

Fig. 2.6 The vector diagram of the beats and the timing diagram for their envelope

$R_b^{-1} \ll R_H C_H$. Then the voltage across the detector load has no time to "follow" the change in the total voltage amplitude $U_{m\Sigma}$ with the beat frequency Ω_b. Since voltages (2.7) are not modulated, the constant voltage U_0 across the detector load can be written in the following form:

$$U_0 = K(U_{m1} + U_{m2}) \tag{2.8}$$

If the voltages u_1, and u_2, supplied to the detector, are respectively tonally modulated in amplitude by the parameters m_1, Ω_1, and m_2, Ω_2, then the voltage across the load of the detector $U_{m_{out}}$ of the envelope amplitudes:

$$U_{m_{out}} = K_D(m_1 U_1 + m_2 U_2). \tag{2.9}$$

Expressions (2.8) and (2.9) allow us to conclude that that the detector, inertial with respect to the beat frequency, has the same transmission coefficient for both input signals, the beat frequency, i.e. $R_b^{-1} \gg R_C C_C$ the condition is fulfilled. To quantitatively evaluate the process of detecting two signals by an inertialess detector, we write down the resulting voltage at its input in accordance with Fig. 2.6, and in the form of a relation

$$u_\Sigma = U_{m\Sigma}(t)\cos[\omega_1 t + \varphi_\Sigma(t)], \tag{2.10}$$

where $U_{m\Sigma}(t)$ is the amplitude of the total voltage, determined from the triangle $0O_1A_1$ by the cosine theorem,

$$U_{m\Sigma(t)} = \sqrt{U_{m1}^2 + U_{m2}^2 + 2U_{m1}U_{m2}\cos(\omega_2 - \omega_1)};$$

$\varphi_\Sigma(t)$—the current phase of the total voltage,

$$\varphi_\Sigma(t) = \mathrm{arctg}\left[\frac{U_{m2}\sin(\omega_2 - \omega_1)t}{U_{m1} + U_{m2}\cos(\omega_2 - \omega_1)t}\right].$$

Thus, the sum of two sinusoidal oscillations of different frequencies can be represented as one oscillation, modulated in amplitude and phase.

Due to the fact that the amplitude detector does not respond to the phase of the detected voltage, in what follows we will take into account only the amplitude of the total voltage. As a result of its detection, a voltage will appear on the load of the inertialess detector, which can be represented as follows:

$$U_{m\,out}(t) = K_D U_{m\Sigma}(t) = K_D U_{m1}\sqrt{1 + x}, \tag{2.11}$$

where $x = \frac{U_{m2}^2}{U_{m1}^2} + 2\frac{U_{m2}}{U_{m1}}\cos(\omega_2 - \omega_1)t$.

For, the inequality $x < 1$.

Then for the expression $\sqrt{x+1}$ we use the binomial expansion

$$\sqrt{x+1} \approx 1 + \frac{1}{2}x - \frac{1}{8}x^2 + \ldots$$

Restricting ourselves to three terms of the expansion (which gives an error in calculations of less than 5%), the expression for the voltage across the detector load can be written as

$$U_{m\,out}(t) = K_D\left[U_{m1} + \frac{1}{4}\frac{U_{m2}^2}{U_{m1}^2} + U_{m2}\cos(\omega_2 - \omega_1)t\right]. \tag{2.12}$$

Analysis of expression (2.12) shows that the action of the second signal during detection manifests itself in two ways. First, on the detector load, in addition to the component linearly related to the amplitude of the larger signal U_{m1}, there appears a component quadratically related to the amplitude of the weak signal U_{m2}. U_{m2}.

Secondly, a component appears on the load that changes with the difference frequency (beat frequency).

Let us consider the obtained result in relation to the cases occurring in the practice of radio reception and noted at the beginning of this subsection [6, 14].

Let the detection of an unmodulated useful signal u_1 and interference u_2, which is not completely filtered out in the MRC, be carried out. In this case, the voltage component with the difference frequency $\omega_2 - \omega_1$ (beat frequency Ω_b) in formula (2.12) plays the role of interference, which can be heard in the form of an unwanted whistle. If the beat frequency Ω_b is higher than the modulating frequency of the signal, then the voltage components at the output of the detector with frequencies Ω_b, $2\Omega_b$ can be quite simply filtered using, for example, a low-pass filter.

The constant component on the detector load is as follows:

$$U_{0\Sigma} = K_D\left(U_{m1} + \frac{1}{4}\frac{U_{m2}^2}{U_{m1}}\right) = U_{01} + U_{02}, \tag{2.13}$$

where $U_{01} = K_D U_{m1}$, $U_{02} = K_D U_{m2}^2/4U_{m1}$.

Expression (2.13) allows calculating the ratio of voltages U_{01} and U_{02} at the output of the detector:

$$\frac{U_{02}}{U_{01}} = \frac{1}{4}\frac{U_{m2}^2}{U_{m1}^2}. \tag{2.14}$$

Formula (2.14) shows that if the ratio of the interference to the signal at the input of the detector is less than one, then at the output it decreases even more, and vice versa. This circumstance allows us to conclude that the transfer coefficient of the inertialess detector is not the same for strong and weak signals, and it turns out to be large for a strong signal than for a weak one [16].

In other words, in an inertialess detector, a weak signal is suppressed by a strong one. It should be noted that the greater the difference in the amplitudes of the detected oscillations, the more pronounced the suppression effect is. From a physical point of view, this can be explained using Fig. 2.6 as follows. If one unmodulated voltage is applied to the detector's input, then a constant voltage u_1 is generated across its load $U_0 = K_D U_{m1}$. If the second voltage u_2 is applied to the detector input, then beatings occur, as a result of which the law of variation of the amplitudes of the envelope of the total oscillation will differ from the sinusoidal one, i.e. it will become asymmetrical. In this case, at the load of the detector, the constant component will increase by a certain amount and become equal to $U_{0\Sigma} = K_D U_{m1} + \Delta U$. Since the increment ΔU is due only to the asymmetry of the envelope of the total oscillation, ΔU it turns out to be less than $K_D U_{m2}$. This indicates that the transmission coefficient for a weak signal turns out to be lower than for a strong one.

A similar result is obtained in the case of detecting two amplitude-modulated signals. Substituting the amplitudes of the envelopes of the detected signals into formula (2.14) and assuming that the modulation coefficients are small, one can obtain

$$\frac{U_{m\Omega 2}}{U_{m\Omega 1}} = \frac{1}{2}\frac{m_2}{m_1}\frac{U_{m2}^2}{U_{m1}^2}, \tag{2.15}$$

where $U_{m\Omega 1}$ and $U_{m\Omega 2}$ are the low-frequency voltages at the detector output, varying with the modulation frequency of the strong and weak signals.

When receiving amplitude-keyed telegraph signals by ear, as noted in Sect. 2.1, in addition to the received signal, vibrations from a special local oscillator are supplied to the amplitude detector, which differs in frequency from the signal by about 1000 Hz. A useful result of detection in this case is a component that changes with a difference frequency $\omega_2 - \omega_1$, which, after filtering and amplification, can be monitored, for example, in telephones. It should be noted that in order to improve the quality of reception, the amplitude of the local oscillator voltage in this case is usually much higher than the amplitude of the received signal; therefore, relation (2.12) turns out to be valid [17].

2.4 Synchronous Detector

One of the main disadvantages of an amplitude detector is the suppression of weak signals by strong ones. This drawback is eliminated in the synchronous detector proposed by the Soviet scientist B. G. Momot in 1934.

The simplest synchronous detector is a multiplier in the form of a nonlinear or parametric element, to which, along with the useful signal $u_C = U_{mC}(t)\cos(\omega_C t + \varphi_C)$,, a voltage $u_G = U_{mG}\cos\omega_{mG} t$ from a local oscillator is supplied, and a low-pass filter (LPF). The heterodyne voltage is synchronized with the signal voltage, i.e. $\omega_G = \omega_C$.

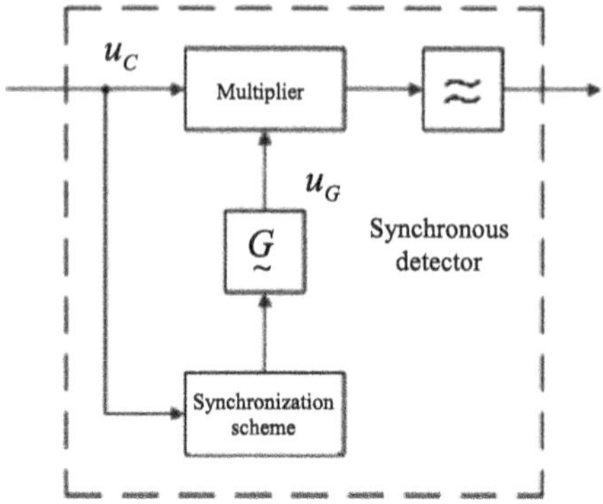

Fig. 2.7 Synchronization circuit

Therefore, such a detector is called synchronous. To ensure the condition of synchronization in the circuits of synchronous detectors, various synchronization circuits are used (Fig. 2.7).

Since (as will be shown below) the mixer plays the role of a multiplier in frequency converters, any frequency converter with a low-pass filter can play the role of a synchronous detector.

Taking into account relations (2.14) and (2.12), we write the expression for the current at the output of the synchronous detector:

$$i_{out} = Y_{21if} U_{mC}(t) \cos \varphi_C. \tag{2.16}$$

Assuming thc dctector load resistance for baseband frequencies to be purely active, we write the expression for the voltage at the detector output:

$$i_{out} = Y_{21if} U_{mC}(t) R \cos \varphi_C. \tag{2.17}$$

This expression shows that the voltage at the output of the detector really depends on the voltage at its input. However, the difference between a synchronous detector and a conventional detector is that the voltage at its output depends on the phase difference $\varphi_C - \varphi_G$ between the received signal and the local oscillator. (In this case $\varphi_\Gamma = 0$). Therefore, such a detector can provide phase separation of the signal. From relation (2.17) it follows that the voltage at the output of the detector reaches its maximum value at $\varphi_C = 0$, i.e. when the phases of the detected signal and the reference voltage of the local oscillator are equal [18].

$$U_{m\Omega} = Y_{21if} m U_{m0} R. \tag{2.18}$$

In this case, the transmission coefficient of a synchronous detector can be determined by the formula

$$K_D = \frac{U_{m\Omega}}{m U_{m0}} = Y_{21if} R. \tag{2.19}$$

Expression (2.19) shows that the transmission coefficient of a synchronous detector K_D does not depend on the amplitude of the received signal and, therefore, there is no suppression of a weak signal in the detector.

From a physical point of view, the latter circumstance can be explained by the fact that oscillations of the local oscillator, adding up with the oscillations of the received signal due to their in-phase, increase the signal-to-noise ratio at the detector input to a value exceeding unity. All other interfering signals with frequencies different from the local oscillator frequency create beats with it, which can be filtered out using appropriate filters.

However, it should be noted that the implementation of the advantages of the synchronous detector is difficult due to the complexity of obtaining the synchronous voltage of the local oscillator. There are three ways to obtain a synchronous voltage [9, 17]:

- By isolating the carrier oscillation of the received signal and amplifying it to the required value;
- By capturing the oscillations of the local oscillator with a dedicated carrier wave;
- By auto-tuning the local oscillator to the selected carrier wave with phase accuracy.

If the in-phase oscillations of the local oscillator and the carrier of the received signal cannot be ensured, then a quadrature synchronous detector is used (Fig. 2.8).

In this detector, the received signal is fed to two multipliers, to the second inputs which oscillations from the local oscillator are supplied u'_G and u''_G, are shifted relative to each other in-phase by 90°

$$\left.\begin{aligned} u'_G &= U_{mG}\cos\omega_G t = U_{mG}\cos\omega_C t, \\ u''_G &= U_{mG}\sin\omega_G t = U_{mG}\sin\omega_C t. \end{aligned}\right\} \tag{2.20}$$

In accordance with the relation (2.17), the voltages at the outputs of the low-pass filters can be written in the form

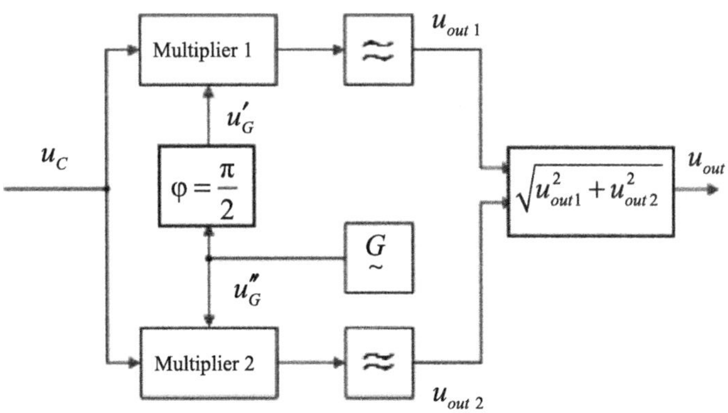

Fig. 2.8 A quadrature synchronous detector

$$\left.\begin{aligned} u_{out\,1} &= Y_{21if} U_{mC}(t) \cos \varphi_C, \\ u_{out\,2} &= Y_{21if} U_{mC}(t) \sin \varphi_C. \end{aligned}\right\} \tag{2.21}$$

The voltages (2.21) from the outputs of the low-pass filters are applied to the quadrator, where, after appropriate transformations, the voltage at the output of the detector is obtained

$$U_{out} = \sqrt{U_{out\,1}^2 + U_{out\,2}^2} = Y_{21if} U_{mC}(t) R \tag{2.22}$$

Thus, the output voltage of the quadrature detector is proportional to the amplitude of the input signal and does not depend on the phase difference between the oscillations of the local oscillator and the signal.

2.5 Distortion of Amplitude-Modulated Signals During Detection

When detecting AM signal along with useful nonlinear transformation of the signal associated with the release of a low-frequency voltage that changes according to the modulation law of the input high-frequency voltage, signal distortions are also possible. The detector can contain both nonlinear signal distortions and linear ones.

It should be noted that the primary cause of nonlinear signal distortions during detection is the presence of a nonlinear element. This leads to the fact that such distortions can arise due to the following reasons [19]:

- Nonlinearity of the detector characteristic;
- Inertia (complexity) of the detector load;
- Complexity of the load (presence of a transition chain on next stage).

Let's consider in more detail these distortions and ways to reduce them.

1. Distortion due to nonlinearity of the detector response. As you know, the real detector characteristic differs from the ideal one and has two characteristic sections: nonlinear and linear (Fig. 2.9). Let's designate the length of the nonlinear section of this characteristic through U_N (Fig. 2.9a). Let the signal presented in the form of relation (2.1) arrive at the input of the detector. When working on a linear section of the detector characteristic, the law of change I_0 in time exactly repeats the law of change in the envelope of the input oscillation (Fig. 2.9b and c, curve I). When working on a nonlinear section, nonlinear distortions will appear (dashed curve 2).

In order for the nonlinear distortions due to the considered cause to be absent, the condition must be met

$$U_{mmin} \geq U_N. \tag{2.23}$$

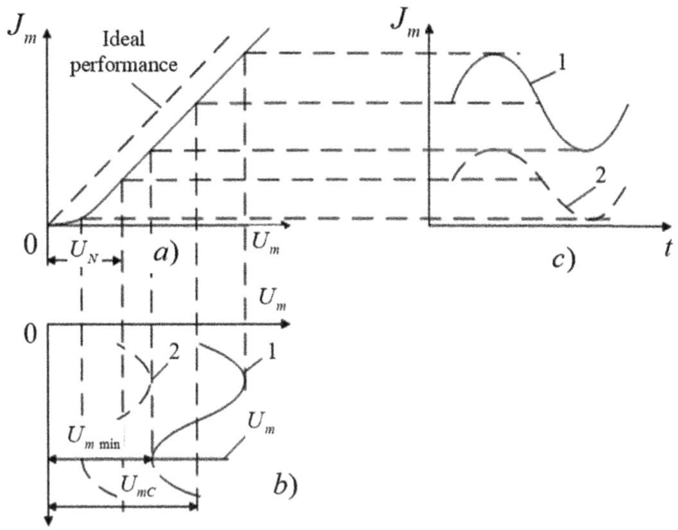

Fig. 2.9 The real detector characteristic

Taking into account the amplitude of the carrier vibration U_{mH} and the maximum modulation coefficient $m_{\max}$, condition (2.23) can be written in the form [20]

$$U_{mC}(1 - m_{\max}) \geq U_N. \tag{2.24}$$

From here

$$U_{mC} \geq \frac{U_N}{(1 - m_{\max})}. \tag{2.25}$$

Usually, the nonlinear section of the detector characteristic corresponds to voltages $U_N \leq (0,1 \div 0,2)$ V.

Assuming that $m_{\max} = 0,9$, from formula (2.25), we obtain that the amplitude of the carrier oscillation U_{mC} should be 1–2 V to eliminate nonlinear distortions.

2. Distortion due to the inertia of the detector load. To clarify the nature of nonlinear distortions caused by this cause, and the conditions for their elimination, we will use the circuit shown in Fig. 2.10.

Consider the timing diagrams of the input high-frequency voltage with sinusoidal modulation and the corresponding voltages across the detector load with different time constants (Fig. 2.11).

As a result of a series of charges and discharges of the capacitor C_1, a low-frequency voltage is generated across the detector load U_{out}.

If the components of the detector load C_1 and R_1 are selected in such a way that for any parameters of the envelope of the input signal U_m, the rate of change in

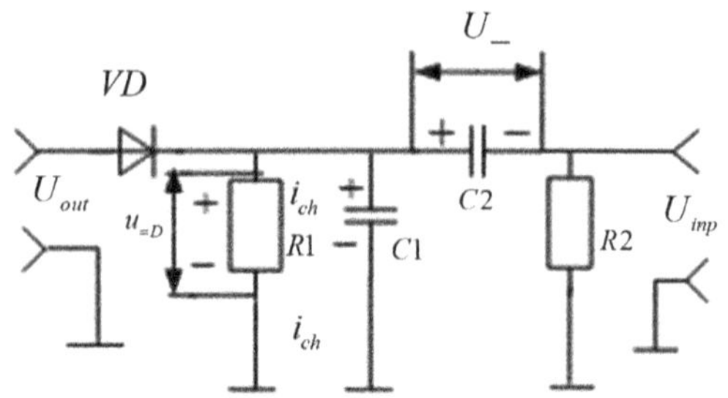

Fig. 2.10 The circuit, using for clarification of the nature of nonlinear distortions

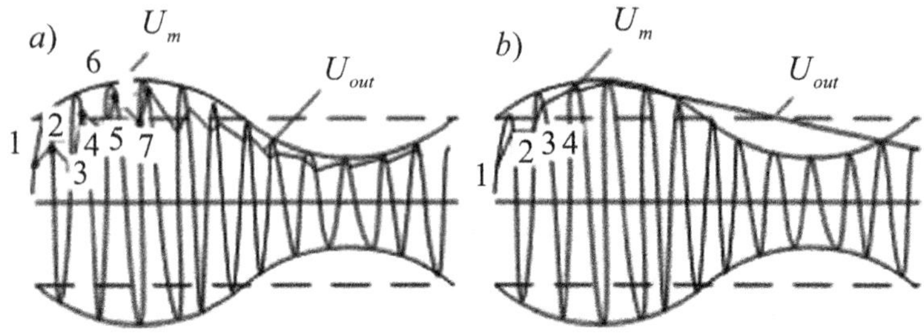

Fig. 2.11 The timing diagrams of the input high-frequency voltage with sinusoidal modulation and the corresponding voltages across the detector load with different time constants

the voltage at the output of the detector exceeds the rate of change in the envelope of the input high-frequency voltage, then the signal distortion, as can be seen from Fig. 2.11a, absent. If this condition is not met, then starting from a certain moment in time, as can be seen from Fig. 2.11b, the capacitor practically does not have time to discharge during the time between adjacent half-periods of the input voltage. In this case, the low-frequency voltage across the detector load does not repeat the shape of the envelope of the input high-frequency voltage, which leads to nonlinear signal distortions during detection. Thus, in order for these distortions to be absent, the condition must be met

$$\left|\frac{dU_{out}}{dt}\right| \geq \left|\frac{dU_m}{dt}\right|. \tag{2.26}$$

For a quantitative assessment of the conditions for the absence of distortions caused by the inertia of the load, consider the fulfillment of inequality (2.26) in relation to point 6 (Fig. 2.11b). Let us write the expression for the envelope of the input signal in accordance with formula (2.1) in the following form [21]:

$$U_m = U_{mH}[1 + m\cos(\Omega t + \varphi)]. \tag{2.27}$$

In addition, we will assume that at the moment of time corresponding to point 6, the voltage across the capacitor is equal $U_{out\ 1}$, and the capacitor is discharged exponentially, i.e.

$$U_{out} = U_{out\ 1} \exp\left(-\frac{t}{R_1 C_1}\right). \tag{2.28}$$

Taking derivatives of expressions (2.27) and (2.28) and substituting them into formula (2.26), we obtain

$$\frac{U_{out\ 1}}{R_1 C_1} \exp\left(-\frac{t}{R_1 C_1}\right) \geq m\Omega U_{mC} \sin(\Omega t + \varphi). \tag{2.29}$$

To strengthen the calculations, we will assume that $t = 0$. In addition, we take into account that at point 6 at the moment when the diode $U_{out\ 1} = U_m$ is turned off, i.e. in accordance with expression (2.27) $U_{out\ 1} = U_{mC}(1 + m \cos\varphi)$. Therefore, we write the relation (2.29) in the following form [22]:

$$\frac{1}{R_1 C_1} \geq \left| \frac{m\Omega \sin\varphi}{1 + m \cos\varphi} \right|. \tag{2.30}$$

The resulting inequality should be fulfilled with the most unfavorable location of point 6 on the envelope, i.e. at any phase value φ. The most unfavorable is the phase value at which the right-hand side of inequality (2.30) is maximum.

Examining the extremum of the right-hand side of inequality (2.30), finding that it is maximal at $\cos\varphi = -m$ and equal to $\frac{m\Omega}{\sqrt{1+m^2}}$.

Therefore, the condition for the absence of distortion takes the form

$$\Omega R_1 C_1 \leq \frac{\sqrt{1 - m^2}}{m}. \tag{2.31}$$

When modulating with a set of baseband frequencies with a variable modulation factor, it is necessary to substitute the upper modulating frequency $\Omega_{\max}$ and the maximum modulation factor $m_{\max}$ in the formula (2.31).

Thus, it can be seen from formula (2.31) that nonlinear distortions caused by the inertia of the load will manifest themselves to a lesser extent both with a decrease in the detector load $R_1 C_1$ time constant and with a decrease in the modulation coefficient and frequency.

3. Distortion due to the complexity of the load. To consider the nonlinear distortions caused by this cause, and the conditions for their elimination, we use Fig. 2.10. Here, the elements R_2, C_2 represent a transition chain to the next stage, while the capacitor C_2 separates the detector and amplifier for direct current.

When the voltage described by expression (2.1) is applied to the detector, the blocking capacitor in the steady state is charged to a constant voltage across the detector load

$$U_0 = U_{mC} \cos\theta. \tag{2.32}$$

When the input voltage decreases during modulation to a minimum value, the capacitor becomes a constant voltage source. It is discharged through the resistors R_1 and R_2, creating an additional voltage U_{0D} across the resistor R_1, which can be determined as follows:

$$U_{0D} = \frac{U_0}{R_1 + R_2} R_1 \tag{2.33}$$

The voltage U_{0D} is applied by "plus" to the cathode of the diode VD, and "minus" to its anode, thereby blocking it. Therefore, when the amplitude U_m of the detected voltage exceeds the voltage U_{0D}, the diode VD is open and, during the detection process, the voltage across the load of the detector repeats the law of change in the envelope of the input oscillation.

If the amplitude of the input oscillation voltage U_m becomes less than the voltage U_{0D}, then the diode VD remains closed, the output voltage of the detector is cut off (Fig. 2.12) and, therefore, nonlinear signal distortions occur. In order for there to be no nonlinear signal distortion in this case, the condition must be met [23]

$$U_{m\ \min} \geq U_{0D}. \tag{2.34}$$

Assuming that $\cos\theta = K_D = 1$ and $U_{m\ \min} = U_{mC}(1 - m)$, taking into account formula (2.32), inequality (2.34) can be written in the form

$$m \leq \frac{R_1}{R_1 + R_2} \text{ or } R_2 \geq \frac{m}{1 - m} R_1. \tag{2.35}$$

It can be seen from relations (2.35) that in order to eliminate nonlinear distortions caused by the presence of a transition chain, it is necessary either to decrease the modulation coefficient m of the detected signal, or to decrease the load resistance of the detector R_1, or to increase the resistance R_2.

However, it should be noted that a decrease in the load resistance of the detector R_1 leads, on the one hand, to a decrease in its transfer coefficient, and on the other, to a decrease in the input impedance of the detector and, consequently, to an increase

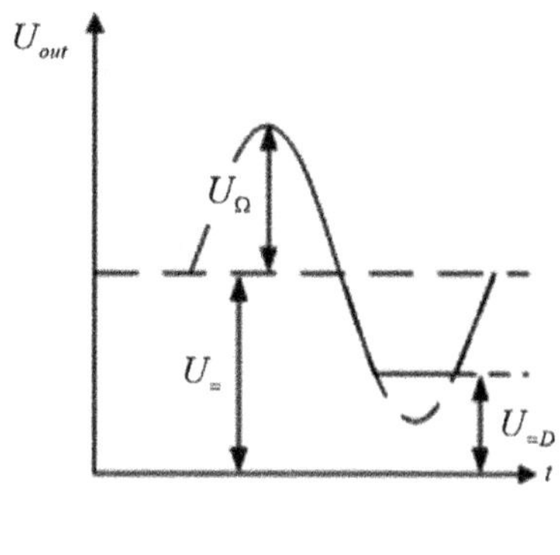

Fig. 2.12 The case, if the timing diagrams of the input high-frequency voltage with sinusoidal modulation and the corresponding voltages across the detector load with different time constants

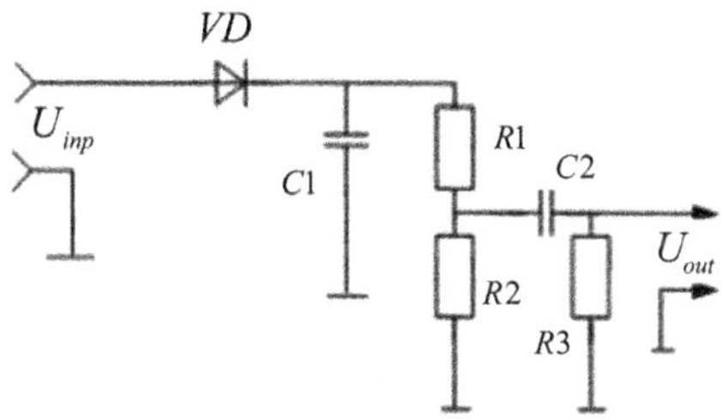

Fig. 2.13 The detector circuit with a divided load

in its shunting effect on the circuit of the previous stage, which in some cases is unacceptable. An increase in the resistance R_2 of the resistor is also not always advisable, since it stands at the input of the stage following the detector and affects its operation. Therefore, in those cases when, in order to eliminate signal distortions, it is not possible to resolve the contradictions that have arisen by choosing the quantities R_1 and R_2, they go to the detector circuit with a divided load (Fig. 2.13).

In this case, the input resistance of the detector is determined by the sum of the resistances R_1 and R_2 and distortion depends only on the magnitude of the resistor R_2. It should be noted that the transmission coefficient of the considered circuit will be less than in the circuit with an undivided load, and it is determined by the formula

$$K_{D_{SL}} = K_D \frac{R_2}{R_1 + R_2}, \tag{2.36}$$

where K_D is the transmission coefficient of the detector at a constant load.

4. Amplitude-frequency distortion of the signal during detection. These distortions are manifested in the fact that the gain of the detector for different baseband frequencies will be different.

The dependence of the detector K_D transmission coefficient on the modulation frequency Ω is shown in Fig. 2.14. It can be seen from this figure that the gain of the detector has a roll-off in the region of the upper and lower modulating frequencies. To find out the reasons for these distortions, we will use the circuit shown in Fig. 2.10. The frequency-dependent elements of this circuit are the capacitors C_1 and C_2, which affect the gain of the detector. In this case, the capacitor C_1 affects the decrease in the transfer coefficient in the region of the upper modulating frequencies, and the capacitor C_2—in the region of the modulating frequencies.

Indeed, the impedance of the detector load can be written as the following relationship:

$$Z_L = \frac{R_1}{1 + j\Omega C_1 R_1}. \tag{2.37}$$

With increasing frequency Ω, the resistance of the capacitor C_1 will decrease. At a certain frequency, the resistance of the capacitor C_1 will be so small that it will begin to shunt the resistance R_1, and this will lead to a decrease in the transmission

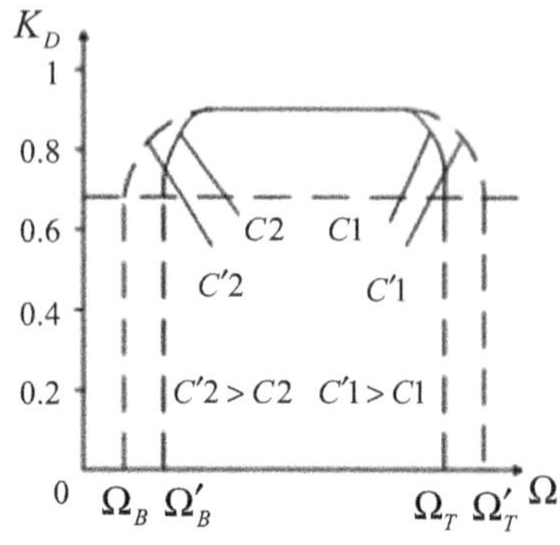

Fig. 2.14 The dependence of the detector K_D transmission coefficient on the modulation frequency Ω

coefficient of the detector. The frequency at which the gain decreases to 0.7 of its maximum is called the cutoff frequency. Distinguish between lower Ω_B and Ω_T upper cutoff frequencies. To increase the upper cutoff frequency, it is generally necessary to decrease the load time constant $\tau_L = C_1 R_1$. In particular, if the resistance R_1 is left unchanged, then to increase the upper cutoff frequency $\left(\Omega'_T > \Omega_T\right)$, it is necessary to decrease the value C'_1 (Fig. 2.14, dashed curve).

In the region of the lower modulating frequencies, the chain C_1, R_1 has practically no effect on K_D, since the resistance of the capacitor C_1 in this case is high and it does not shunt the resistor R_1.

In the region of lower modulating frequencies, on the contrary, the decay of the detector transmission coefficient is influenced by the separation circuit C_2, $R2$ the impedance Z_I of which can be written in the form [24]

$$Z_I = \frac{j\Omega C_2 R_2 + 1}{j\Omega C_2}. \tag{2.38}$$

At the upper modulating frequencies, the resistance of the capacitor C_2 is small and it has practically no effect on their passage to the next stage. All the low-frequency voltage taken from the detector load is then dropped across the resistor R_2. At lower modulating frequencies, with a decrease in the modulation frequency, the resistance of the capacitor C_2 to these frequencies increases, therefore, the voltage drop across it increases, and across the resistor R_2, it decreases, which leads to a decrease in the detector transmission coefficient. In this case, to decrease the lower cutoff frequency, it is necessary to increase the time constant of the transient circuit $\tau = C_2 R_2 . \tau = C_2 R_2$.

In particular, if you leave the resistance R_2 value unchanged, then you can reduce the lower cutoff frequency $\left(\Omega'_B < \Omega_B\right)$ by increasing the capacitance of the capacitor C_2 (see Fig. 2.14).

2.6 Features of Signal Receivers with One Side Strip

As noted in Sect. 2.1, the most efficient from the energy point of view and from the point of view of occupied bandwidth, SSB signals are signals with weakened carrier ($A3A$) and completely suppressed carrier ($A3J$). The total voltage gain in signal-to-noise ratio during the transition from AM signal to SSB signal does not exceed 4 (12 dB).

Since single-sideband (SSB) signals are a modification of the well-known AM signal, in general the linear path of SSB receivers has much in common with the linear path of AM signal receivers. These receivers are usually superheterodyne. Due to the fact that signals with SSB type $A3A$ and $A3J$ have a spectral width that is two times smaller than the spectrum width of the corresponding AM signal, then the bandwidth of the linear path of an SSB receiver is two times narrower than that of AM signal receivers. In this case, the high-frequency channel of the receiver must be tuned so that the carrier oscillations are at the border of the passband [5, 9–12].

Reception of signals from SSB, in the general case, imposes higher requirements on the linearity of the MRP, on the stability of its amplitude-frequency characteristic, and on selectivity. However, the main feature of SSB receivers, which distinguishes them from AM signal receivers, is mainly determined by the method of detection by devices for implementing this method. To provide detection, such receivers have a recovery or carrier shaping system. The need for such a system is explained by the following circumstance. When the amplitude detector detects the voltage of one side band (for example, the upper one) $U_{USS} = \frac{mU_{mC}}{2} \cos(\omega_C + \Omega)t$, only a constant voltage will be released on the load, while it is necessary to obtain a low-frequency voltage, follow the modulation law. This is possible only when, simultaneously with the SSB signal, a voltage with the carrier frequency is applied to the detector ω_C, i.e. voltage $u = U_{mC} \cos \omega_C t$. In this case, as a result of the detection of beats that occur between the components of the SSB signal and the carrier, low-frequency components appear corresponding to the modulation frequencies.

The voltage of the carrier oscillation, which must be supplied to the detector, can be obtained from the following sources [20, 24]:

- From a local high-stability generator;
- From a controlled local oscillator, the frequency of which automatically changes in accordance with the frequency of the filtered and amplified incompletely suppressed carrier or in accordance with a special pilot signal, the frequency of which is selected outside the spectrum of the modulation side frequencies and differs from the frequency of the carrier vibration by a predetermined amount.

When using the pilot signal, it is additionally converted to the frequency of the carrier wave. It should be noted that the pilot or the remainder of the carrier in the receiver is used to ensure the normal operation of the AGC and AFC systems.

Single-sideband signal detection can be performed by two methods:

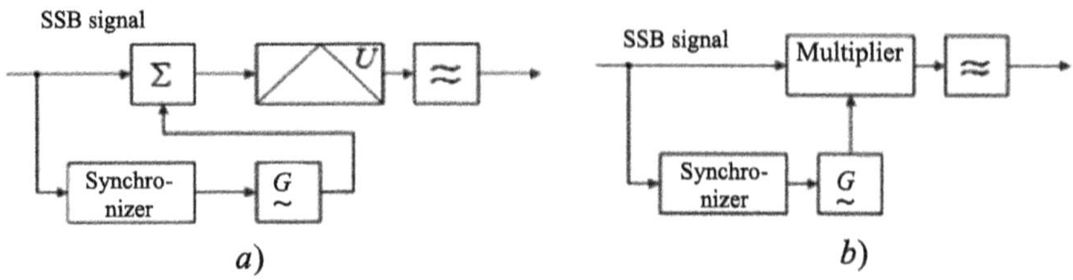

Fig. 2.15 Controlled local oscillator: **a** heterodyne detection; **b** frequency converter and phase detector.

- Using an amplitude detector, to which the SSB signal is fed directly through the adder and the carrier restored by one of the above methods, for example, using a (heterodyne detection, Fig. 2.15a);
- Using a multiplier, to which the SSB signal and oscillations of the local oscillator with a frequency equal to the carrier frequency are also fed. The role of the multiplier can be performed by any frequency converter and phase detector (Fig. 2.15b).

When detecting by the first method, beats occur as a result of the summation of the SSB signal and oscillations of the carrier oscillator. As shown in Sect. 2.3, the beating frequency of the envelope is equal to the difference of the summed oscillations and, therefore, is equal to the modulation frequency, which is released at the load of the amplitude detector [21].

It should be noted that in the heterodyne detection of a SSB signal, the same distortions may occur as in the detection of a conventional AM signal, discussed in Sect. 2.5. In addition, when detecting an SSB signal, two new types of distortions are possible. These distortions are caused by:

- Insufficient voltage amplitude of the local or recovered carrier;
- Insufficient synchronism between local and suppressed carrier oscillations.

Indeed, as noted in Sect. 2.3, the shape of the beating envelope curve at comparable amplitudes of the detected oscillations is non-sinusoidal and asymmetric, which is the reason for the occurrence of nonlinear distortions. To reduce such distortions during detection, it is necessary to work with a large amplitude of the local oscillator voltage. The relationship that establishes the relationship between the amplitude of the local or recovered carrier U_{mC}, the amplitude of the SSB $U_{m\,SSB}$ signal and the harmonic coefficient K_G, which characterizes the permissible level of nonlinear distortion, can be written as

$$\frac{U_{mC}}{U_{m\,SSB}} \approx \frac{1}{4K_G} \tag{2.39}$$

So, if the permissible value of the harmonic distortion $K_\Gamma = 2.5\%$, then the required voltage amplitude U_{mC} in accordance with expression (2.39) must exceed the voltage amplitude $U_{m\,SSB}$ by 10 times.

When detecting a single-sideband signal using the second method in accordance with the diagram shown in Fig. 2.15b, a single-sideband signal and oscillations of the local oscillator are fed to the mixing element. With a sufficiently large amplitude of oscillations of the local oscillator, the slope of the mixing element changes with the frequency of the local oscillator, as a result of which, as indicated in Sect. 6.1, combination components of the total and difference frequencies appear. In this case, the difference frequency is equal to the modulation frequency of the SSB signal and is allocated using a low-pass filter [15, 17].

With the second detection method, nonlinear signal distortions may occur due to the nonlinearity of the slope characteristic of the mixing element.

In addition to the distortions of the SSB signal during detection in the SSB receiver, distortions are also possible due to an error in the recovery of the carrier frequency. The inaccuracy of carrier recovery is manifested in the fact that all components of the baseband spectrum (Ω_1, Ω_2, Ω_3 etc.) receive the same increment during detection, equal to the recovery error, which leads to sound distortion.

For high quality telephone reception, the allowable carrier recovery asynchronism should not exceed 30 Hz.

In conclusion, let us give as an example one of the possible schemes of a signal receiver with SSB (Fig. 2.16).

The SSB signal, having passed the preselector and the IF amplifier, is separated by a sideband filter. The FSSB should have a frequency response with a high slope to suppress all frequency components not belonging to the SSB signal. To realize good.

FSSB selectivity, the intermediate frequency must be relatively low; therefore, in such receivers, in some cases, multiple frequency conversion is used. A narrow-band filter, the bandwidth of which does not exceed 50 Hz, allocates the remainder of the carrier. This signal is used to operate the AGC system and AFC of the first local oscillator, and also synchronize the frequency of the generator of the local carrier using the phase-locked frequency system (PLF), which includes a phase detector (PD), a low-pass filter and a control element (CE). A single-sideband signal from

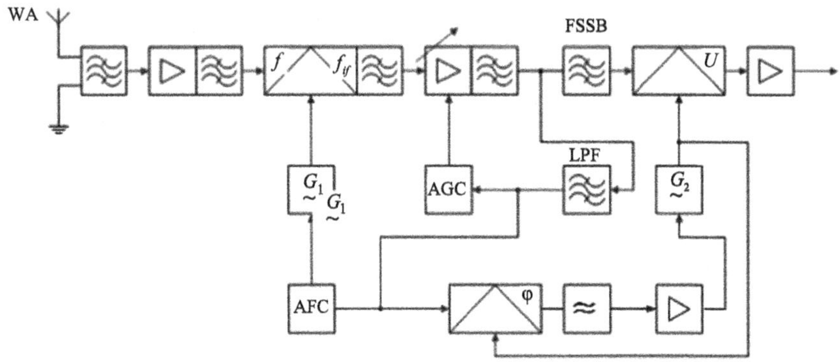

Fig. 2.16 A signal receiver with SSB

the output of the FSSB and oscillations with the carrier frequency from the local oscillator are fed for detection to an amplitude detector, from the output of which the signal is fed through an ultrasonic signal to the terminal device [23].

References

1. Van der Ziel (1973) Noise (sources, description, measurements). Translated from English. M: Sov. radio, p 178
2. Voishvillo GV (1983) Amplifying devices. M: Radio and communication, p 264
3. Vorobiev VI (1983) Optical location for radio engineers. M: Radio and communication, p 176
4. Gutkin LS, Chentsova OS (1958) Transient processes in the system: high-frequency amplifier—detector. Radio engineering, no. 11, pp 18–23
5. Ditkin VA, Prudnikov AP (1965) A guide to operational calculus. M: Higher. shk., p 466
6. Evtyanov SI (1948) Transient processes in receiving-amplifying circuits. M: Radio and communication, p 210
7. Zernov NV, Karpov VG (1972) Theory of radio engineering circuits. Leningrad: Energiya, p 816
8. Vasilchenko NV, Borisov VA, Kremenchug LS, Levin GE (1983) Measurement of parameters of optical radiation receivers. In: Kurbatova LN, Vasilchenko NV (eds) M.: Radio and communication, p 320
9. Kanevsky ZL, Finkelstein MI (1963) Fluctuation interference and detection of pulsed radio signals. M; L: Gosenergoizdat, p 243
10. McElroy JH (1977) Communication systems for near space using CO_2 lasers. TIIER, no. 2, pp 54–89
11. Malashin MS, Kaminsky RP, Borisov YB (1983) Basics of designing laser locating systems: Textbook manual for universities—M: Higher. shk., p 207
12. Martynov VA, Selikhov YI (1980) Pan-frame receivers and spectrum analyzers. In. Zavarin GD (ed) M: Sov. radio, p 352
13. Pervachev SV (1982) Radioautomatics: Textbook. for universities. M: Radio and communication, p 296
14. Pestryakov VB, Kuzenkov VD (1985) Radio engineering systems. M: Radio and communication, p 376
15. Poberezhsky ES (1987) Digital radio receivers. M: Radio and communication, p 184
16. Prat VK (1972) Laser communication systems: Translated from English M: Svyaz, p 232
17. Golubkov AP, et al. (1984) Design of radar receiving devices: Textbook for radio engineering special universities. In: Sokolov MA (ed) M: Higher school, p 335
18. Bankov VN, Barulin LG, Zhodzishsky MI, et al. (1984) Radio receivers. In: Barulina LG (ed) M: Radio and communication, p 272
19. Zyuko AG, Korobov YF, Levitan GI, Simontov IM, Falco AI (1975) In: Zyuko AG (ed) Radio receivers. M: Communication, p 400
20. Teplyakov IM, Roshchin BV, Fomin AI, Weizel VA (1982) Radio systems for transmitting information: Textbook for universities. In: Teplyakova IM (ed) M: Radio and communication, p 264
21. Ross M (1969) In: Nevsky AV (ed) Laser receivers: Translated from English, Moscow: Mir, p 520
22. X yu z RS (1980) Logarithmic amplifiers. Moscow: Sov. radio, p 352
23. Zhodzishsky MI, Sila-Novitsky SY, Prasolov VA, et al. (1980) Digital phase synchronization systems. In: Zhodzishsky MI (ed) M: Soviet radio, p 208
24. Chistyakov NI (1938) To the calculation of one discriminator circuit. News of the electrical industry of a weak current, no. 2

Chapter 3
Radio Receivers of Pulse Signals

3.1 Brief Information on Pulse Signals and Features of the Structural Diagrams of Receivers

Pulse signals are widely used in radar and radio navigation, in radio telemetry, in radio communication systems and in a number of other areas of radio engineering. The main parameters of pulse signals are pulse-amplitude U_P; pulse duration τ_P; repetition period T_P (repetition frequency F_P), time position of the beginning of the pulse (pulse-phase).

In the general case, the parameters of pulse signals depend on the purpose of the system and the tactical and technical requirements for it. So, for example, the repetition period in the radar system is selected from the condition of unambiguous measurement of the range to the target, and the pulse duration τ_P—from the condition of the required range resolution [1].

In radio communication and telemetry systems, both discrete and continuous messages are transmitted using pulse signals. The possibility of transmitting a continuous message in short pulses is based on the Kotelnikov theorem. In this case, the duration of the impulses τ_P giving instantaneous values of the continuous message turns out to be much less than the period of their counting. This circumstance makes it possible to place the pulses of another message in the intervals between the pulses of one message, and thus to carry out multichannel communication.

By changing one of the parameters of the pulse signal in accordance with the modulating function, various types of pulse modulation are obtained, namely, pulse-amplitude modulation (PAM), pulse-width modulation (PWM), pulse-phase (time) modulation (PPM), pulse-code modulation (PCM), etc. [2].

To ensure the required tactical and technical characteristics of the systems listed above, pulses are used, the duration of which ranges from tenths of a microsecond to units—tens of microseconds. Such short durations of pulse signals indicate their wide spectrum.

S. V. Dvornikov et al., *Radio-Electronic Equipment in Civil Aviation*, Springer Aerospace Technology, https://doi.org/10.1007/978-981-19-6199-1_3

The real width of the spectrum of a pulse signal, in which the main part of the energy is concentrated, depends on its duration, shape and is determined using the relation

$$\Delta f_S = \frac{a}{\tau_P}, \tag{3.1}$$

where a is a coefficient that varies depending on the pulse shape.

With an accuracy sufficient for practice for any pulse shape, its spectrum width can be determined by the formula

$$\Delta f_S = \frac{1}{\tau_P}. \tag{3.2}$$

If increased requirements are imposed on undistorted pulse transmission, then the width of its spectrum is found using the expression

$$\Delta f_S \approx \frac{1}{\tau_l}, \tag{3.3}$$

where τ_l is the settling time of the leading edge of the pulse ($\tau_l < \tau_P$). Taking into account formulas (3.2) and (3.3), the spectrum of pulses used in modern systems can range from several hundred kilohertz to units or even tens of megahertz.

Depending on the purpose of the system, pulse signal receivers can operate in the meter, decimeter and centimeter wavelength ranges. It should be noted that, in the general case, the operating frequency range significantly affects both the structural diagram of the receiver as a whole and the choice of appropriate amplifying devices and types of resonant systems [3].

To meet the high requirements for sensitivity and selectivity for modern pulse signal receivers, they are built according to the superheterodyne circuit. Depending on the purpose of the system, pulse signal receivers can be single-channel or multichannel. So, the receivers of radar stations are most often single-channel, and the receivers of communication radio relay stations are multichannel. Figure 3.1 shows, respectively, simplified block diagrams of a single-channel receiver and a multichannel time division receiver.

As you can see from these diagrams, the receivers have many common elements. Let us give a brief description of some of the elements of these circuits, emphasizing their features and taking into account the structure of the signal.

The main reception path, as in any receiver, determines the sensitivity of the receiver, its single-signal and multi-signal selectivity, and ensures the normal operation of the detector. Since the bandwidth of the receiver is selected in accordance with the width of the spectrum of the received signals, taking into account the above, the receivers of pulse signals are relatively wideband, which in some cases makes it possible to build them with a single frequency conversion [4].

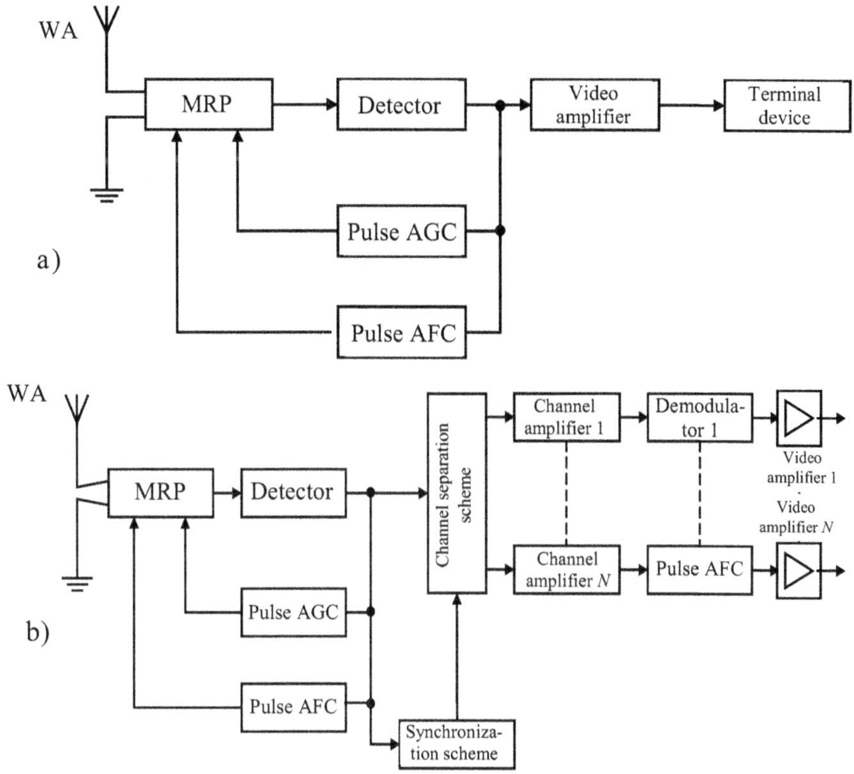

Fig. 3.1 Simplified block diagrams of a single-channel receiver and a multichannel time division receiver

To form a wide bandwidth of the receiver and to provide high gain as the IF amplifier of such receivers, amplifiers with twos and threes of detuned circuits are usually used.

A carrier detector in a single-channel circuit converts a high-frequency pulse signal into video pulses, and in a multichannel circuit—into a group signal while maintaining the primary modulation law.

The channel separation and pulse synchronization scheme (PSS) provides distribution of the baseband signal pulses to individual channels in accordance with the sequence formed in the transmitter. For this purpose, marker (synchronizing) pulses allocated according to a certain parameter are used, with the help of which the so-called dummy pulses for individual channels are formed, which open the corresponding channel only at the moment of arrival of the pulse of this channel. After separation into channels and appropriate amplification of the pulses, they are demodulated using channel detectors to extract information. Additional amplification to the level required for the normal operation of the terminal device is provided by video amplifiers (VA) [3, 5].

Maintaining a constant voltage at the output of the receiver with a significant change in the level of input signals is ensured by a pulse AGC system, and stable reception of signals with random drifts of the transmitter and receiver local oscillator frequency is achieved using a pulsed AFC system. The peculiarities of the operation of these systems and their structural diagrams will be considered in Sect. 3.4.

It should be noted that in the range of meter and decimeter waves, the spectral width of the received pulse signal is much greater than the possible frequency drifts of the transmitter and local oscillator due to their instability, as well as the maximum possible Doppler shift. In this case, the expansion coefficient turns out to be close to unity, and therefore the use of the AFC system in such receivers in most cases turns out to be inappropriate.

3.2 Distortion of Pulse Signals in Electoral Amplifiers

When pulsed signals pass through the selective amplifiers of the receiver, transients occur in them, which leads to signal distortion. To consider the issue of the passage of pulse signals through selective amplifiers, we will use the method of superposition or transient characteristics, the essence of which is that the input signal of a complex shape is represented as a sum of elementary signals, for example, voltage surges (single jumps). By finding the transient processes caused by the action of all elementary signals that form a complex signal, and summing up the useful results, one can find the desired transient process and, therefore, estimate the shape of the pulses at the output of the selective amplifiers. Let us use the above to analyze the passage of video pulses and radio pulses through selective amplifiers [6].

I. Passage of video pulses through selective amplifiers. Suppose that a video pulse with an amplitude equal to one and a duration is supplied to a single-stage amplifier τ_P (Fig. 3.2a). The supply of such a video pulse to the amplifier is equivalent to the effect on it of two voltage surges (Fig. 3.2c), equal in magnitude, opposite in sign and shifted relative to each other by time τ_P.

The transient response of the electoral circuit is an oscillating function.

Let us take the moment of the beginning of the pulse as the origin of time. Then the voltage at the amplifier output, caused by the application of the first voltage jump at the moment of time, can be written as follows:

$$u_{\text{out}}(t > 0) = K_0[A(t)\sin\omega_0 t]. \quad (3.4)$$

After applying the second voltage jump ($t = t + \tau_P$), which corresponds to the end of the input pulse, the output voltage of the selective circuit has the form:

$$u_{\text{out}}(t \geq t + \tau_P) = K_0\{A(t)\sin\omega_0 t - A(t + \tau_P)\sin\omega_0(t + \tau_P)\}, \quad (3.5)$$

where

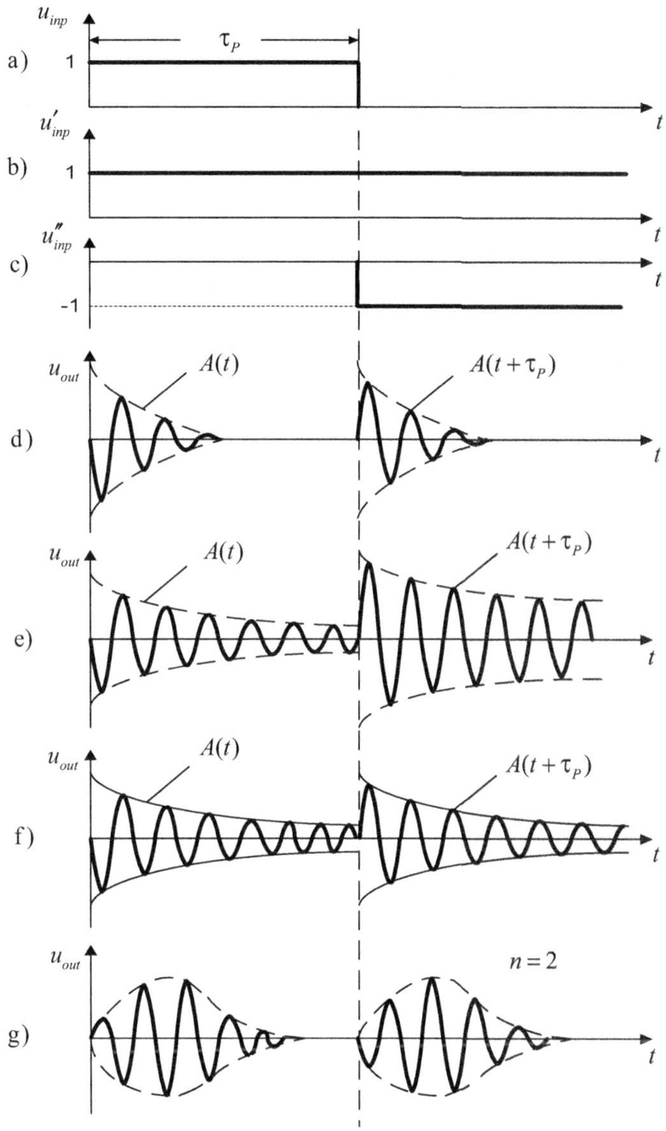

Fig. 3.2 Passage of video pulses through selective amplifiers

$$U_{\text{out}}(t) = K_0\sqrt{A^2(t) + A^2(t+\tau_P) - 2A(t)A(t+\tau_P)\cos\omega_0\tau_P} \tag{3.6}$$

is the voltage envelope at the amplifier output, and the initial phase of the envelope

$$\text{tg}\varphi(t) = \frac{A(t)\sin\omega_0\tau_P}{A(t)\cos\omega_0\tau_P - A(t+\tau_P)}. \tag{3.7}$$

In accordance with expression (3.6), the envelope of the voltage at the output of the amplifier depends not only on the envelope of its transient characteristics, but also on the phase relations of natural oscillations at the moment of applying the second switch-on function, which is taken into account by the term [7]

$$2A(t)A(t+\tau_P)\cos\omega_0\tau_P.$$

Figure 3.2d shows the timing diagrams of the voltage at the output of a single-stage resonant amplifier for the case when the pulse duration τ_P is much longer than the settling time τ_P, therefore, the voltage at the output arising from the first jump completely decays by the moment $t{+}\tau_P$. In this case, in the circuit of the selective amplifier, natural oscillations arise twice—at the moments of the beginning and end of the pulse, which corresponds to the moments when two voltage surges are applied. The envelopes of the output voltages in both cases are equal to each other, i.e.,

$$U_{out}(t) = K_0A(t+\tau_P) = K_0A(t)$$

and change exponentially over time. The phases of the high-frequency filling, on the contrary, are opposite to each other and depend on the phase of the jump of the corresponding switching functions.

Figure 3.2d–f shows the timing diagrams of voltages at the output of a single-stage selective amplifier for the case when the settling time of transient processes τ_l is longer than the pulse duration τ_P. In this case, Fig. 3.2d corresponds to the case when $\cos\omega_0\tau_P = -1$, i.e., when the natural oscillations from the first voltage surge coincide in phase with the natural oscillations that arise at the moment of the second voltage surge. Figure 3.2e, corresponds to the value $\cos\omega_0\tau_P = +1$ when the natural oscillations from these voltage surges are antiphase. In all other cases, the output voltage envelope occupies an intermediate position between the envelopes corresponding to the values $\cos\omega_0\tau_P = \pm 1$.

Thus, when a video pulse is applied to a single-stage selective amplifier, damped oscillations occur at the amplifier tuning frequency in its output circuit. The rate of decrease in the amplitude of these oscillations is determined by the losses in the circuit, and, consequently, by its quality factor. The higher the Q-factor of the contour (less attenuation), the slower the envelope changes $A(t)$ from period to period of high-frequency filling, and vice versa.

If a video pulse is fed to the input of an n-stage amplifier, then the voltage at its output (namely, the amplitude envelope and the phase of the high-frequency oscillation) depends on the number of stages. In this case, the phase of the high-frequency filling coincides with the phase of the jump of the corresponding voltage jump with an even number of stages and is opposite to it with an odd number of stages. In Fig. 3.2g, as an example, the timing diagrams of the voltage at the output of the two-stage selective amplifier are shown for the case when the settling time of transient processes is less τ_P.

II. Passage of radio pulses through selective amplifiers. Suppose that the amplifier receives a radio pulse with a unit amplitude, with a duration τ_P (Fig. 3.3a) and a frequency ω_0 that coincides with the average tuning frequency of the amplifier. The supply of such a radio pulse to the input of the selective amplifier is equivalent to the inclusion of two harmonic antiphase voltages at its input, shifted by the time of the pulse duration τ_P (Fig. 3.3c). Taking, as in the previous case, the moment of the beginning of the pulse as the starting point of time, we write the voltage at the amplifier output caused by the switching on of the harmonic voltage at the moment in time $t = 0$, in the following form:

$$u_{\text{out}}(t > 0) = K_0 B(t) \sin(\omega_0 t + \varphi), \tag{3.8}$$

where $K_0 B(t) = U_{\text{out}}(t)$ is the envelope of the output voltage for $t = 0$ (Fig. 3.3g).

In the same way, when applying the second harmonic voltage corresponding to the end of the radio pulse, the voltage at the amplifier output can be written as the following expression:

$$u_{\text{out}}(t > t + \tau_P) = -K_0 B(t + \tau_P) \sin[\omega_0 (t + \tau_P) + \varphi], \tag{3.9}$$

where $K_0 B(t + \tau_P)$ is the envelope of the output voltage for a point in time $t + \tau_P$ (Fig. 3.3e). The minus sign in the formula (3.9) indicates that the output voltage caused by the application of a voltage at the moment $t = t + \tau_p$ is in the antiphase with respect to the first.

The total voltage at the output of the selective amplifier can be represented by the relation

$$u_{\text{out}} \sum (t \geq t + \tau_P) = K_0 [B(t) - B(t + \tau_P)] \sin[\omega_0 (t) + \varphi]. \tag{3.10}$$

In this case, the envelope of the total voltage at the amplifier output changes according to the law

$$U_{\text{out}} \sum (t \geq t + \tau_P) = K_0 [B(t) - B(t + \tau_P)]. \tag{3.11}$$

If the duration of the radio pulse at the amplifier input is much longer than the settling time, then for $t \geq 0$ can be practically considered $B(t + \tau) \approx 1$. In this case, the envelope of the total output voltage after the end of the pulse in accordance with formula (3.11) has the form

$$U_{\text{out}} \sum (t \geq t + \tau_P) = K_0 [1 - B(t + \tau_P)]. \tag{3.12}$$

Such an envelope is shown in Fig. 3.3e. Taking this into account, a radio pulse is obtained at the output of the selective amplifier, which has the form shown in Fig. 3.3g. As can be seen from this figure, the shape of the radio pulse at the output of the single-stage amplifier is significantly distorted in comparison with the shape

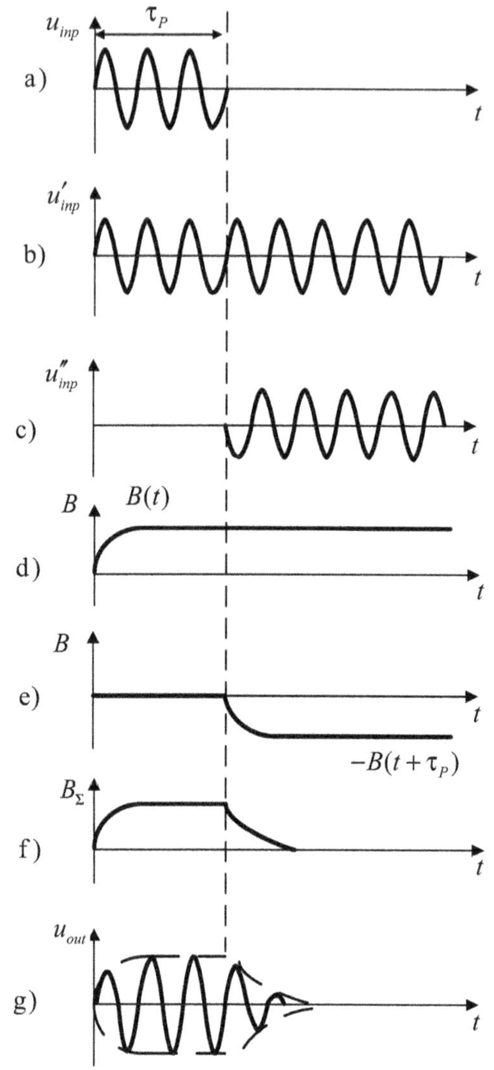

Fig. 3.3 Passage of radio pulses through selective amplifiers

of the input pulse. The distortion of the output radio pulse significantly depends on the ratio of the signal spectrum width (pulse duration) and the amplifier bandwidth. It should also be noted that the distortion of the pulse at the amplifier output will largely depend on the accuracy of the amplifier tuning to the frequency of the received signal [8, 9]. If the frequencies of the received signal and the settings of the amplifier do not coincide with each other, then the envelope of the radio pulse at the output of the amplifier will change with a beat frequency equal to the difference between the above frequencies. The oscillatory nature of the envelope of the radio pulse will also

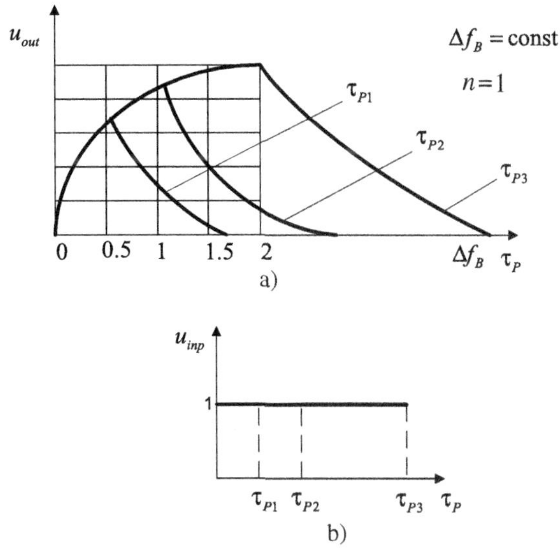

Fig. 3.4 **a** The dependence of the output pulse shape on the product $\Delta f_B \tau_P$; **b** the input of a single-stage resonant amplifier

be in amplifiers with detuned circuits. In this case, the frequency of the envelope oscillations depends on the detuning parameter.

Figure 3.4a shows the dependence of the output pulse shape on the product $\Delta f_B \tau_P$ when a rectangular radio pulse of different duration is applied to the input of a single-stage resonant amplifier (Fig. 3.4b) with a filling frequency equal to the tuning frequency of the amplifier. The construction was made for three values of the product: $\Delta f_B \tau_P$ 0.5; I; 2.

Analysis of the curves (Fig. 3.4a) shows that with a constant bandwidth of the amplifier, an increase τ_P to a value approximately corresponding $1/\Delta f_B$ to it leads to an increase in the amplitude of the output voltage. When the voltage amplitude $\tau_P > 1/\Delta f_B$ at the amplifier output becomes almost equal to its steady-state value.

3.3 Optimization of the Main Tract of Reception of Pulse Signals

It is known that the main operations performed on any received signal at optimal reception are reduced to finding the likelihood ratio and comparing this ratio with the threshold on the solver. When the threshold is exceeded, a decision is made on the presence of a signal, in the opposite case—on the absence of a signal.

In practice, it is not the likelihood ratio itself that is most often sought, but functions monotonically associated with this ratio. One of these functions, in particular, is the correlation integral, which can be written as follows [2]:

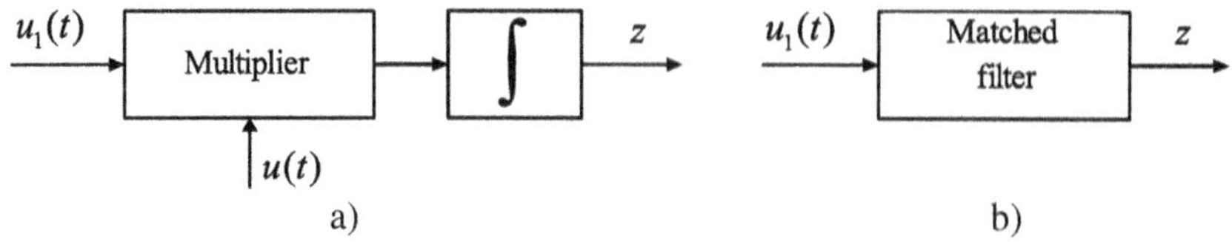

Fig. 3.5 Two methods of reception: **a** the correlation method of reception; **b** the filter method of reception

$$z = \frac{2}{N_0} \int_0^{T_C} u(t)u_1(t)\mathrm{d}t \tag{3.13}$$

where N_0 is interference spectral density; T_C observation interval; $u(t)$ useful signal received; $u_1(t)$ mixture of a useful signal and interference, $u_1(t) = u(t) + n(t)$.

Depending on the method used for calculating the correlation integral (3.13), two methods of reception are distinguished: correlation and filter (Fig. 3.5).

Figure 3.5a shows a block diagram of the correlation method of reception, which displays the operations (3.13) that must be performed on the input mixture. At the input of the receiver, in addition to a copy of the signal, the receiver performs multiplication and integration of the product. The main feature of this technique is that it is not invariant with respect to the delay, i.e., if the signal delay is unknown, then for its detection, it is necessary to build a multichannel receiver based on the delay.

The correlation integral (3.13) can be obtained by another method, namely, by passing a mixture of a signal with noise through a linear filter (Fig. 3.5b), the impulse response of which is a mirror image of the signal. Such a filter is called a matched filter. Its transmission coefficient is a complex conjugate function of the signal spectrum. When the input mixture $u_1(t)$ passes through such a filter, a correlation integral is also obtained at its output. In this case, all the spectral components of the signal at the filter output are added in phase [10, 11].

As a result, the signal-to-noise ratio at the output is equal to $\sqrt{2E/N_0}$, where E is the signal energy. This ratio turns out to be greater than the output of any other filter. The synthesis of the matched filter circuit for the received signal is carried out either according to the transmission coefficient or according to the impulse response, either according to the transmission coefficient or according to the impulse response, taking into account their physical feasibility.

As an example, consider the structural diagrams of matched filters for the simplest pulse signals. Figure 3.6 shows one of the possible matched filter schemes for a rectangular video pulse and timing diagrams of the voltages at the output of its elements. The filter consists of an integrator, a delay line and a subtractor.

The input signal $u(t)$ goes to the integrator. The voltage $u'(t)$ from the integrator output is supplied to the subtractor directly and through the delay line for the duration of the video pulse $u''(t)$. As a result, the voltage at the output of the subtractor will correspond to the output of the matched filter.

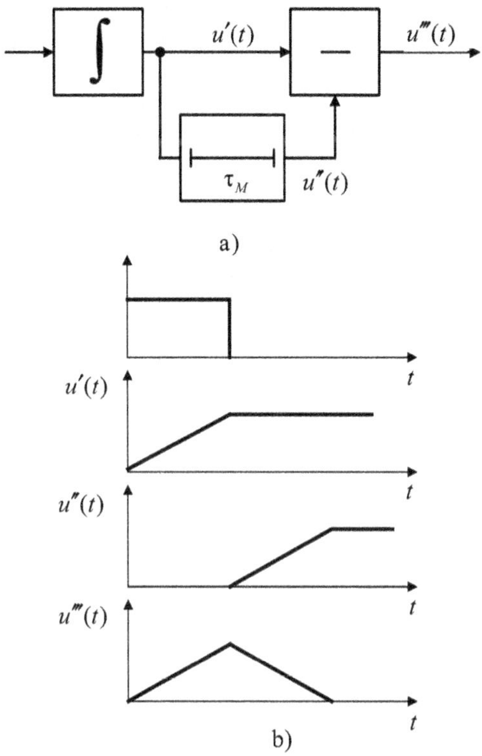

Fig. 3.6 The possible matched filter schemes for a rectangular video pulse and timing diagrams of the voltages at the output of its elements

Similarly, in Fig. 3.7a, b, respectively, the matched filter circuit for a rectangular radio pulse and the voltage across its elements are shown.

In the circuit, the role of the integrating link is played by an oscillatory circuit tuned to the carrier frequency of the radio pulse.

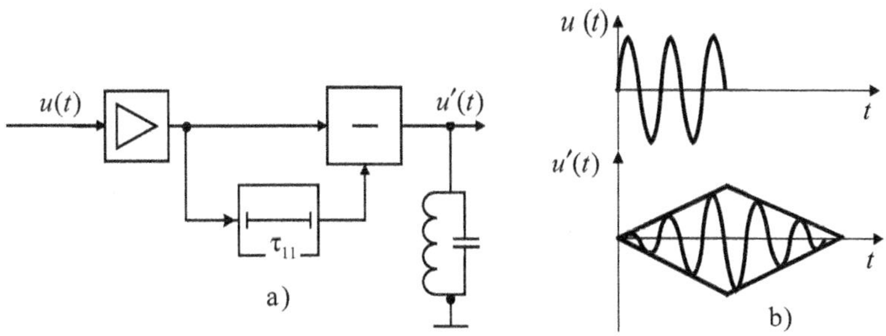

Fig. 3.7 **a** The matched filter circuit for a rectangular radio pulse; **b** the voltage across its elements

However, it should be noted that the implementation of matched filters, especially for signals of a more complex structure and various types of modulation causes great difficulties. Therefore, in practice, instead of optimal filters, quasi-optimal filters are often used, the role of which in the receiver is played by selective amplifiers of the intermediate frequency path, which form the required bandwidth and the required form of the amplitude-frequency characteristic. In this case, the choice of the bandwidth of the pulse signal receiver is usually carried out taking into account the problems solved by a particular radio engineering system as a whole, and the requirements imposed directly on the receiver [4, 12].

So, for example, in receivers, where the main task is to detect and receive weak signals, and the pulse shape does not matter, to ensure the highest sensitivity, it is necessary to obtain the maximum ratio of signal power to the power of intrinsic noise.

The noise power at the output of a selective amplifier is proportional to its bandwidth:

$$P_{N\text{ out}} = 1.1\Delta f_B kT(Noize - 1)K_{PH} \tag{3.14}$$

Therefore, to reduce the influence of intrinsic noise, the bandwidth of the receiver's linear path should be reduced. However, as the bandwidth decreases the oscillation settling time increases. This can lead, as can be seen from Fig. 3.4, to the fact that the amplitude of the pulse signal at the output during the pulse τ_P duration will not have time to increase to its maximum steady-state value. Figure 3.8 shows a graph of the dependence of the signal-to-noise ratio at the output of the resonant amplifier on the bandwidth Δf_B at a constant duration of the input pulse τ_P. Analysis of the curves, shown in Fig. 3.8, shows that there is some optimal value of the amplifier bandwidth $\Delta f_{N\text{ opt}}$, at which the signal-to-noise ratio at its output is maximum.

It can be shown that for any amplifier circuit and any form of a radio pulse acting at its input, formula (3.1) is valid. In this case, the value α of the coefficient depends on the type of the amplitude-frequency characteristic and the type of radio pulse. Table 3.1 shows the values for various combinations of the pulse shape and frequency response of the quasi-optimal filters, as well as the signal-to-noise ratio of

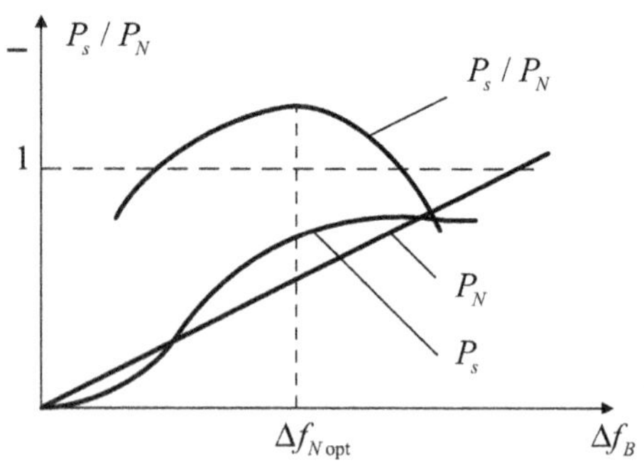

Fig. 3.8 A graph of the dependence of the signal-to-noise ratio at the output of the resonant amplifier on the bandwidth Δf_B at a constant duration of the input pulse τ_P

Table 3.1 The values for various combinations of the pulse shape and frequency response of the quasi-optimal filters

Radio pulse	Filter	a	$\gamma_{K\ \mathrm{opt}}/\gamma_{\mathrm{opt}}$
Rectangular	Rectangular	1.37	0.91
Rectangular	Gaussian	0.72	0.94
Gaussian	Rectangular	0.72	0.94
Gaussian	Gaussian	0.63	1
Rectangular	Single resonant circuit	0.4	0.9
Rectangular	Two-stage resonant amplifier	0.61	0.94
Rectangular Gaussian	Five -stage resonant amplifier	0.67	0.94

the voltage at the output of these filters in comparison with the signal-to-noise ratio for the optimal filter $\gamma_{K\ \mathrm{opt}}/\gamma_{\mathrm{opt}}$.

Analysis of the data, shown in Table 3.1, shows that the value of the optimal bandwidth from the point of view of obtaining the maximum signal-to-noise ratio is not particularly critical. This circumstance is used when it is important to preserve the pulse shape, for example, in the receivers of precision guidance radars, high-precision navigation systems, etc. In this case, the bandwidth is chosen more than the optimal one and it can be approximately calculated by the formula (3.3). However, this causes some deterioration in the signal-to-noise ratio at the output of the receiver and, therefore, decreases the sensitivity [13].

3.4 Features of Automatic Regulation Systems in Receivers of Pulse Signals

Automatic control systems in pulse signal receivers include pulse AFC and AGC systems. The main feature of these systems is the discreteness of the regulating voltages, due to the impulse nature of the received signals. Let us dwell in more detail on the features of these systems. Pulse AFC systems. These systems are divided into two main groups: inertial and high-speed.

In the inertial AFC system (AFCS), the receiver detuning value is stored during the pulse duration, and in the intervals between the pulses, the local oscillator is adjusted. The duration of the transient process in such a system turns out to be longer than the repetition period [14].

In fast-acting AFC systems, the local oscillator frequency is controlled during the pulse duration, therefore, the transient time in this case should be less than the pulse duration. This greatly complicates such systems.

It should be emphasized that both inertial and high-speed pulse AFC systems are most often systems that provide a constant intermediate frequency.

Figure 3.9 shows the AFCS diagram. Let us give a brief description of the features of some of its elements. At the output of the frequency detector (FD), which acts as

Fig. 3.9 AFCS diagram

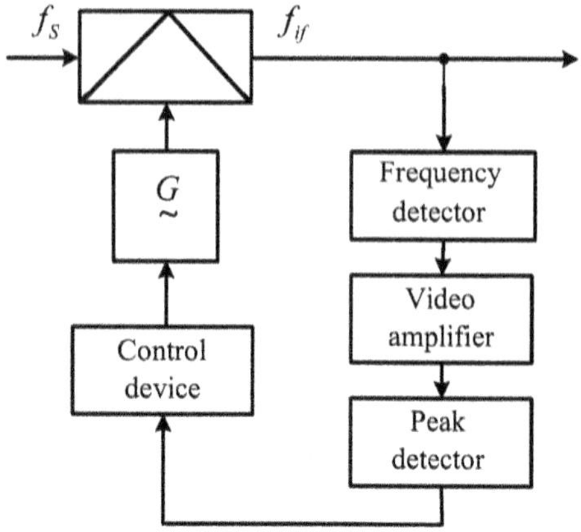

a measuring element, video pulses appear, the amplitude and polarity of which are proportional to the magnitude and sign of the deviation of the intermediate frequency from the nominal value. The duration and repetition period of these pulses correspond to the duration and repetition period of the received signals.

The FD passband in the pulsed mode of operation is determined by the duration. And in practice, it is chosen two to four times the width of the spectrum of the received signal. In this case, the steepness of the FD characteristic turns out to be relatively small, which leads to the need for additional amplification in the control circuit. However, the pulsed nature of the voltage at the FD output makes it possible to relatively simply use a video amplifier to carry out the necessary amplification in the control circuit without resorting to amplifying the constant control voltage, which eliminates the disadvantages associated with direct current amplification [10].

The main difference between a pulse AFCS and a continuous one is in the principle of constructing a filter for a control circuit. The filter circuit must ensure the formation of a constant control voltage proportional to the deviation of the intermediate frequency from the nominal value, i.e., proportional to the amplitude of video pulses at the FD output. In the overwhelming majority of cases, such a filter is a peak video pulse detector, the voltage at the output of which is proportional to the amplitude of the video pulses from the FD output. This voltage is the control voltage and provides, with the help of the controller, the appropriate adjustment of the local oscillator frequency in the intervals between the received pulses.

To increase the reliability of operation, pulsed inertial AFC systems can have an electronic or electromechanical frequency search (EFS) circuit.

Let us dwell on some of the features of the EFS systems. It is known that an increase in the speed of any automatic control system can be achieved by reducing the duration of the transient process, for example, by reducing the time constants of the elements of the control circuit. However, there are limits to this reduction. So, when receiving pulsed signals, the duration of which is less than 1–2 μs, it becomes a very difficult task to ensure the normal operation of the auto-tuning system.

Another equally important factor limiting the capabilities of EFS systems is the signal delay in the stages of the main receive path, as a result of which the process of tuning the local oscillator is shifted by the group delay time. This circumstance leads to the fact that EFS systems are used only in radio receivers that provide the reception of pulsed signals of a relatively long duration. In addition, it is advisable to use the AFC systems in the receivers of radar stations, in which the AFC system can operate not according to the signal reflected from the target, but according to the emitted signal of the transmitter. In this case, the signal level at the input of the AFC system is sufficiently large, which makes it possible to significantly simplify the channel of the automatic tuning of the local oscillator frequency. It should be noted that a number of special measures are used to increase the performance of such EFS systems, however, the circuit solutions in this case turn out to be very complex. Therefore, when receiving pulsed signals of short duration, not exceeding 3–5 μs, it is more expedient to use AFC systems in receivers operating on the principle of stabilizing the local oscillator frequency with an increase in the stability of the transmitter frequency.

AGC systems of pulse receivers. According to the speed of regulation, such systems are subdivided into inertial AGC (IAGC), high-speed AGC (HAGC) and instantaneous AGC (INAGC) [8].

The regulating voltage in the IAGC system is determined by a certain averaged value of the amplitude of the received signals for a relatively large number of pulse repetition periods. In the HAGC system, the regulating voltage is determined by the value of the amplitude of the pulses received for each repetition period. In IAGC systems, the control voltage is determined by the instantaneous value of the received signal level within the pulse duration.

Figure 3.10 shows a block diagram of the IAGC reverse impulse system. In its appearance, it has a complete analogy with the structural diagram of a continuous AGC system, differing from it in some features of the filter circuit of the control circuit. Let us briefly consider these features.

From the output of the pulse detector, video pulses corresponding to the envelope of the input radio pulses are fed, on the one hand, to the processing circuit, and on the other, to the control circuit, where they are converted into a constant regulating

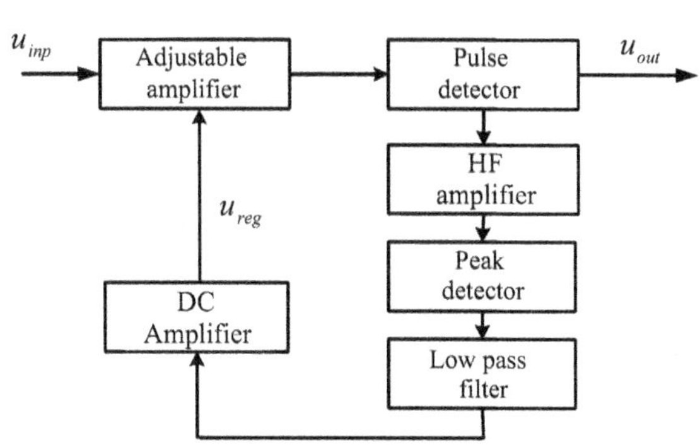

Fig. 3.10 A block diagram of the IAGC reverse impulse system

voltage, in accordance with which either the gain of the adjustable amplifier will change, or the transmission coefficient of the communication circuit between the amplification stages. Such a transformation, as noted when considering the AFC impulse system, can be carried out using a peak detector (video impulse detector), in which the time constant of the discharge of its capacitance is chosen significantly greater than the repetition period T_P ($\tau_{\text{reg}} >> T_P$). To increase the efficiency of the AGC, an additional video amplifier (circled by a dotted line) can be installed in the control circuit.

When receiving signals with PAM, to prevent their demodulation, the filter time constant τ of the IAGC system should significantly exceed the maximum period of the modulated voltage T_M ($\tau \geq (10 \div 20)T_M$). With other types of modulation, for example, PCM, PWM, etc., when the amplitude of the pulses remains unchanged and, therefore, there is no danger of demodulation in the circuit, the speed of the AFCS circuit can be increased. The response speed can also be increased by decreasing the charge and discharge time constants of the video pulse detector. However, this decreases the constant regulating voltage at its output and decreases the AGC efficiency. To improve the efficiency of such systems, as a rule, additional DC amplification is applied.

Large inertia of IAGC systems can cause suppression of received signals by powerful impulse noise. Indeed, the reception of an interference pulse of large amplitude will cause a significant decrease in the receiver gain. Since the gain will recover over a significant period of time, the receiver gain may be insufficient to receive weak wanted signals and the signal will be suppressed. To eliminate this drawback, a noise-immune IAGC is sometimes used in pulse receivers, the structural diagram of which is shown in Fig. 3.11.

In the control circuit (usually between the video amplifier and the peak detector), a delay line (DL) and a key circuit are placed, which are gated by a selector that selects a useful signal. DL provides a signal delay for a time equal to the duration of the selection process. The key circuit passes only useful signals to the input of

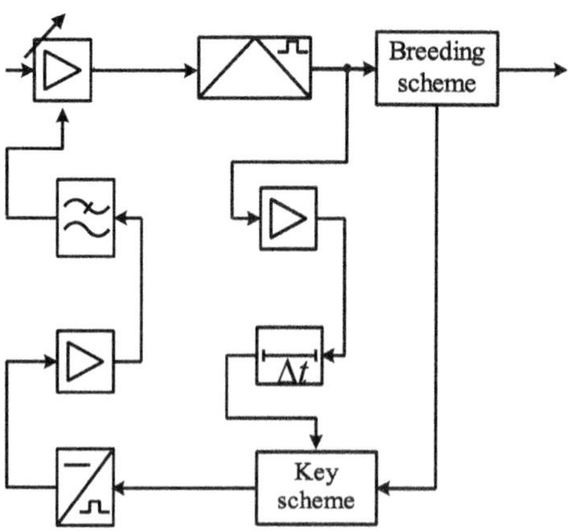

Fig. 3.11 The structural diagram of a noise-immune IAGC

the control circuit, as a result of this, the effect of interference on the AGC circuit at other times is eliminated.

In the HAGC system, the speed can be increased by decreasing the time constants of all elements of the control circuit, mainly the filter time constants. However, such a decrease can lead to the excitation of the receiver. A significant factor limiting the performance of conventional reverse HAGC circuits is the group delay time in the MRP stages. As a result, a strong drop in the top of the received pulses can be observed. To eliminate this drop in HAGC systems, each regulated receiver stage is covered by its own control loop, which eliminates the need for interstage decoupling filters for the control loop. At the same time, their inertia decreases, and, consequently, performance increases. It should be noted that the number of regulated cascades in the receiver, covered by autonomous control circuits, does not exceed two or three [7, 11].

Despite the application of the above measures, the minimum duration HAGC circuit directly controls the receiver gain during the duration of the received pulses only in cases where the pulse duration exceeds 15 μs.

At the same time, HAGC circuits are used in receivers designed to receive pulses of shorter duration. But in this case, the purpose of its application is to protect the receiver from the action of powerful continuous or long-term impulse noise.

IAGC systems are the fastest. It is practically impossible to meet the high requirements for the speed of such systems by the previously considered methods of "backward" adjustment. Therefore, the methods of constructing IAGC systems are currently based on the automatic change in the equivalent resistance, the load of the regulated receiver stages, or the automatic sequential elimination of the receiver stages, which are overloaded as a result of receiving signals with a high level. These methods of constructing IAGC systems allow obtaining the so-called logarithmic amplitude characteristic of the receiver, which provides a logarithmic relationship between the output and input voltages. The principle of constructing logarithmic amplifiers will be discussed later.

References

1. Belousov AP (1959) Calculation of the noise figure of radio receivers. State. publishing house of defense. prom-sti, M., 135p
2. Bogdanovich BM (1980) Nonlinear distortions in receiving and amplifying devices. Svyaz, Moscow, 280p
3. Ditkin VA, Prudnikov AP (1965) A guide to operational calculus. Higher. shk., M., 466p
4. Evtyanov SI (1948) Transient processes in receiving-amplifying circuits. Radio and communication, M., 210p
5. Zernov NV, Karpov VG (1972) Theory of radio engineering circuits. Energiya, Leningrad, 816p
6. Malashin MS, Kaminsky RP, Borisov YuB (1983) Basics of designing laser locating systems: textbook manual for universities. Higher. shk., M., 207p
7. Martynov VA, Selikhov YuI (1980) Pan-frame receivers and spectrum analyzers (Zavarin GD (ed)). Sov. radio, M., 352p

8. Pervachev SV (1982) Radioautomatics: textbook for universities. Radio and communication, M., 296p
9. Pestryakov VB, Kuzenkov VD (1985) Radio engineering systems. Radio and communication, M., 376p
10. Poberezhsky ES (1987) Digital radio receivers. Radio and communication, M., 184p
11. Zyuko AG, Korobov YuF, Levitan GI, Simontov IM, Falco AI (eds) (1975) Radio receivers. Communication, M., 400p
12. Teplyakova IM, Roshchin BV, Fomin AI, Weizel VA, (eds) (1982) Radio systems for transmitting information: textbook for universities. Radio and communication, M., 264p
13. Ross M (1969) Laser receivers (Translated from English) (Nevsky AV (eds)). Mir, Moscow, 520p
14. Sheremetyev AG (1971) Statistical theory of laser communication. Communication, Moscow, 264p

Chapter 4
Radio Receivers of Signals with Angle Modulation

4.1 General Information on Reception of Frequency-Modulated Signals

Frequency modulation (FM) is widely used in telemetry systems, as well as in telephone, radio relay and satellite communication systems. The use of FM significantly improves the use of transmitter power and increases the noise immunity of reception [1].

In general terms, a frequency-modulated signal is described by the expression

$$u_C(t) = U_{mC} \cos\left[\omega_C t + \varphi_C + \Delta\omega_m \int_0^t M_C(\tau)\,\mathrm{d}\tau\right], \quad (4.1)$$

where $M_C(\tau)$ is the modulating function and $\Delta\omega_m$—frequency deviation.

With harmonic modulation $M_C(t) = \cos \Omega t$, and

$$u_C(t) = U_{mC} \cos(\omega_C t + \varphi_C + m_f \sin \Omega t), \quad (4.2)$$

where m_f is the frequency modulation index, $m_\mathrm{f} = \frac{\Delta\omega_\mathrm{C}}{\Omega} = \frac{\Delta f_\mathrm{m}}{F}$ and Ω—modulation frequency $\Omega = 2\pi F$.

The frequency modulation index characterizes the amplitude of the change in the phase of the modulated oscillation. The spectrum of such fluctuations is theoretically infinite. However, in practice, if we restrict ourselves to taking into account the spectral components having amplitudes of at least 10% of the carrier, the spectrum width is determined by the ratio:

$$\Delta f_\mathrm{S} \approx 2Ff\sqrt{m_{f_\mathrm{max}}}. \quad (4.3)$$

Frequency Shift Keying (FSK) is used to transmit discrete messages. In this case, the modulating function $M_C(t) = \Pi_C(t)$ is a sequence of positive and negative pulses of unit amplitude with the duration of the elementary message τ_0. In FSK,

S. V. Dvornikov et al., *Radio-Electronic Equipment in Civil Aviation*, Springer Aerospace Technology, https://doi.org/10.1007/978-981-19-6199-1_4

the deviation of the transmitter center frequency is estimated by the frequency $\pm\Delta f_c$ deviation or frequency shift $2\Delta f_C = f_{2C} - f_{1C} = 2\Delta f_m$, and the modulation index is determined by the ratio of the frequency deviation to the average keying frequency.

The spectrum width of such a signal [2, 3]

$$\Delta f_{BS} \approx \frac{a}{\tau_0} + 2\Delta f_m. \tag{4.4}$$

The value of the coefficient $a \geq 1$ is determined by the permissible degree of distortion of the shape of the elementary message in the receiver path.

The advantages of frequency modulation are most fully realized at high modulation indices ($m_f \gg 1$), which leads to a large width of the spectrum of the received signals. Therefore, mainly frequency-modulated signals are used at decameter and shorter wavelengths.

A characteristic feature of the block diagram of FM receivers is the presence of an amplitude limiter in front of the frequency detector. Since the useful information is contained in the change in the signal frequency, then with the help of the amplitude limiter it is possible to significantly weaken the harmful amplitude modulation of the signal by interference, which makes it possible to improve the quality of signal reception and increase the noise immunity of reception [4]. The realization of these advantages is possible only in the case when the signal-to-noise ratio at the output of the receiver's linear path is $\gamma \geq 2$, and the signal level at the limiter input is several times higher than the limiting threshold.

The detection of frequency-modulated signals is carried out by frequency detectors, the output voltage of which is proportional to the instantaneous value of the frequency of the detected oscillations.

The distortion of the law of change in the instantaneous frequency of the signal in the selective systems of the receiver is the cause of the occurrence of nonlinear distortions during detection. Therefore, in FM receivers, strict requirements are imposed on the linearity of the phase-frequency characteristics of the linear path [5].

Frequency-modulated radio links can be either single channel or multichannel. There are a number of methods for constructing multichannel systems with frequency modulation for the transmission of telegraph, telephone and telemetric signals, which differ mainly in the nature of the construction of their decoding devices used to separate channels and optimize signal processing in them. Otherwise, the block diagram of the FMS receiver repeats the block diagram of the amplitude-modulated signal receiver. Accordingly, all the elements of its receiving path perform similar functions, providing the required sensitivity of the radio receiver, its selectivity for adjacent and side receiving channels, multi-signal selectivity, dynamic range, etc.

4.2 Distortions of Frequency-Manipulated Signals in the Linear Tract of the Receiver

Consider the physical processes occurring in a selective amplifier when a frequency-shift keyed signal passes through it. As shown in Fig. 4.1, the harmonic voltage, undergoing a frequency $2\Delta\omega$ jump at the moment $t = 0$, can be represented in the form of two oscillations:

$$u_1 = \begin{cases} u_{11} = U_{m1}e^{j(\omega_C-\Delta\omega)t}\text{for } t < 0; \\ u_{12} = U_{m1}e^{j(\omega_C+\Delta\omega)t}\text{for } t \geq 0. \end{cases} \quad (4.5)$$

We will assume that the first of these two voltages was applied to the amplifier input for a long time, so that by the time $t = 0$ the transient process caused by the switching on of this voltage has practically ended and a harmonic signal u_{B1} with a frequency exists at the amplifier output $\omega_C - \Delta\omega$. Under these conditions, at $t = 0$ the output of the amplifier, three voltages will arise: natural oscillations due to the end of the voltage u_{C1}, natural oscillations and forced ones u_{C1}, due to the appearance of voltage u_{12}. In this case, natural oscillations u_{C1} and u_{C2} have a frequency ω_C

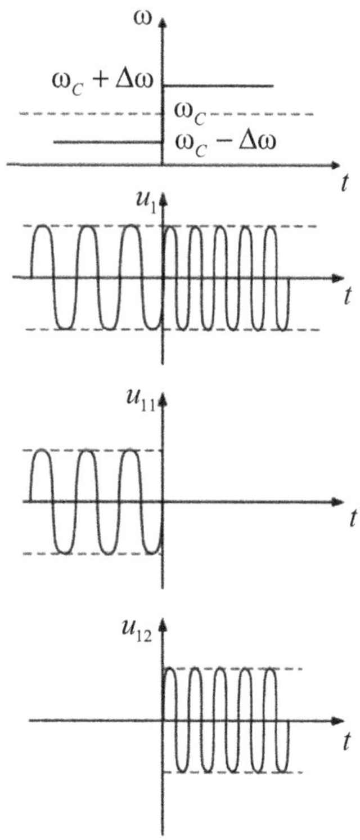

Fig. 4.1 The harmonic voltage, undergoing a frequency $2\Delta\omega$ jump at the moment $t = 0$

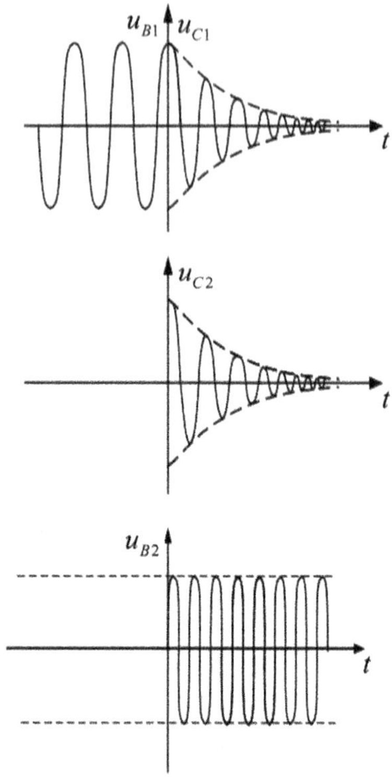

Fig. 4.2 The corresponding timing diagrams of these voltages

corresponding to the tuning frequency of the amplifier, and an amplitude at the time of their occurrence, equal to the amplitude of forced oscillations and decreasing in time according to the law $1 - B(t)$ determined by the transient characteristic of the amplifier amplitude.

The corresponding timing diagrams of these voltages are shown in Fig. 4.2.

When constructing them, it was taken into account that the forced oscillations u_{B2}, due to detuning $\Delta\omega$, passing through the amplifier, receive a phase shift relative to the voltage u_{12} equal to $-\varphi$ (see Fig. 4.3), and also that the natural oscillations u_{C2} at the moment $t = 0$ are antiphase to the forced ones u_{B2}, and the natural oscillations u_{C1} are in phase with the forced ones u_{B1}, which have a phase shift relative to voltage u_{11} equal to φ. The corresponding vector diagrams are shown in Fig. 4.4.

Thus, for $t \geq 0$

$$\left.\begin{aligned} u_{C1} &= U_m[1 - B(t)]e^{j(\omega_C t+\varphi)}, \\ u_{C2} &= U_m[1 - B(t)]e^{j(\omega_C t+\pi-\varphi)}, \\ u_{B1} &= U_m e^{j[(\omega_C+\Delta\omega)t-\varphi]}. \end{aligned}\right\} \tag{4.6}$$

Based on the principle of superposition, the output voltage can be represented as the algebraic sum of these three voltages.

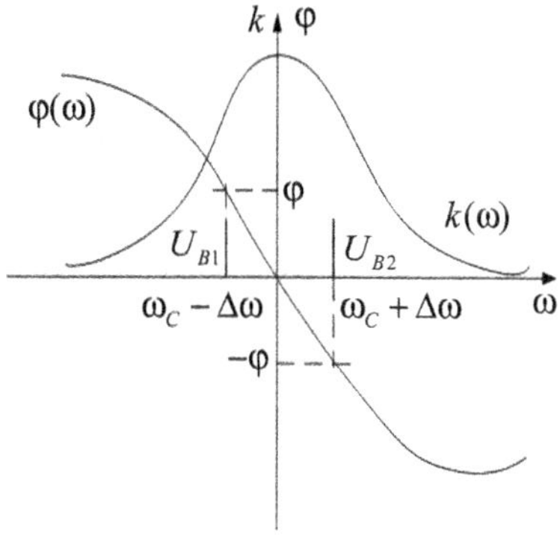

Fig. 4.3 A phase shift relative to the voltage u_{12} equal to $-\varphi$

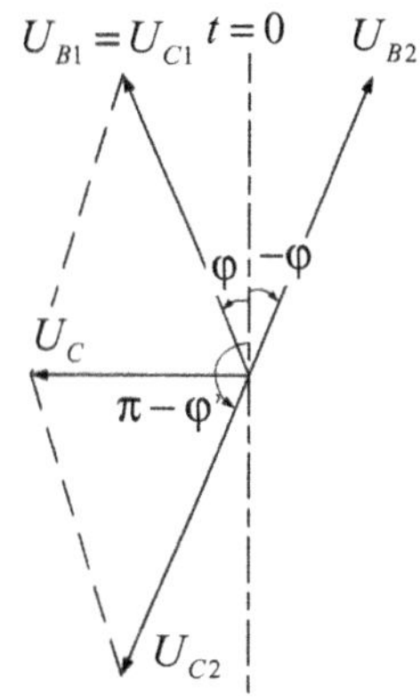

Fig. 4.4 The corresponding vector diagrams

For a single-stage resonant amplifier $B(t) = 1 - \mathrm{e}^{-\alpha t}$, the output voltage at $t \geq 0$ is respectively equal to

$$u_2 = u_{C1} + u_{C2} + u_{B2} = U_m e^{-at} e^{j\omega_C t}\left[e^{j\varphi} + e^{j(\pi-\varphi)}\right] + U_m e^{j[(w_C+\Delta w)t-\varphi]} = \\ = U_m e^{j[(\omega_C-\Delta\omega)t-\varphi]}\left[1 + e^{-\alpha t} e^{-j\Delta\omega t}(e^{j2\varphi} - 1)\right]. \tag{4.7}$$

In relation (4.7), the expression in square brackets is the normalized complex amplitude of the amplifier output voltage at $t \geq 0$:

$$\dot{U}(t \geq 0) = 1 + \mathrm{e}^{-\alpha t}\mathrm{e}^{-j\Delta\omega t}(\mathrm{e}^{j2\varphi} - 1). \tag{4.8}$$

Assuming by Euler's formula

$$\mathrm{e}^{-j\Delta\omega t} = \cos\Delta\omega t - j\sin\Delta\omega t,$$
$$\mathrm{e}^{j2\varphi} = \cos 2\varphi + j\sin 2\varphi,$$

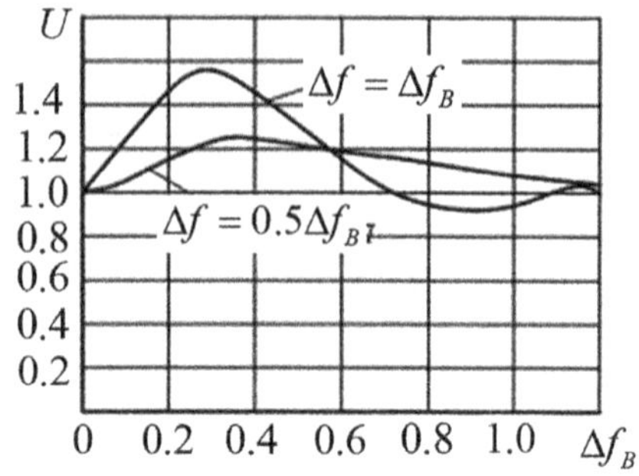

Fig. 4.5 The dependence of voltage on various values of the frequency jump at its input, built on the basis of expression (4.11)

after elementary trigonometric transformations, we get:

$$\begin{aligned}\dot{U}(t \geq 0) &= 1 + e^{-\alpha t}[\cos(\Delta\omega t - 2\varphi) - \cos \Delta t] + \\ &+ je^{-\alpha t}[\sin \Delta\omega t - \sin(\Delta\omega t - 2\varphi)].\end{aligned} \tag{4.9}$$

Expression (4.9) makes it possible to obtain the dependence of the envelope setting and the frequency of the output voltage on time:

$$\left.\begin{aligned} U(t \geq 0) &= \sqrt{[\mathrm{Re}\dot{U}(t)]^2 + [\mathrm{Im}\dot{U}(t)]^2}, \\ \omega(t \geq 0) &= \frac{\mathrm{d}}{\mathrm{d}t}\mathrm{arctg}\frac{\mathrm{Im}\dot{U}(t)}{\mathrm{Re}\dot{U}(t)}. \end{aligned}\right\} \tag{4.10}$$

Thus, from expression (4.9), taking into account (4.10), after simple transformations, we obtain (Fig. 4.5)

$$U(t \geq 0) = \sqrt{1 + 2^{-\alpha t}[\cos(\Delta\omega t - 2\varphi) - \cos \Delta\omega t] + 2e^{2\alpha t}(1 - \cos 2\varphi)}. \tag{4.11}$$

Figure 4.6 shows the graphs of the dependence of the envelope settling at the output of a single-stage resonant amplifier on various values of the frequency jump at its input, built on the basis of expression (4.11), and in Fig. 4.6—similar graphs of the dependence of the frequency settling. An analysis of these curves shows that when the frequency jumps $2\Delta f < \Delta f_B$ (which is always the case in practice), the increase in the amplitude of the output voltage does not exceed 25% of the steady-state value, and the nature of the frequency setting is approximately the same as when the amplitude at the output is set when the input is applied a harmonic signal amplifier with a single hop envelope [6–8].

At the moment of time $t = 0$, the geometric sum of the vectors of the forced $\dot{U}_{B2}$ and natural vibrations $\dot{U}_C = \dot{U}_{C1} + \dot{U}_{C2}$ is equal to $\dot{U}_{B1}$, at $t > 0$, since the frequency of the natural vibrations w_C is less than the forced ones $\omega_C + \Delta\omega$, the vector $\dot{U}_C$, decreasing in amplitude, rotates relative to the vector $\dot{U}_{B2}$ with frequency $\Delta\omega$. As a result, the output voltage of the amplifier $\dot{U}_2 = \dot{U}_C + \dot{U}_{B2}$, the end of the vector of which slides along the dotted curve, changes in amplitude with the

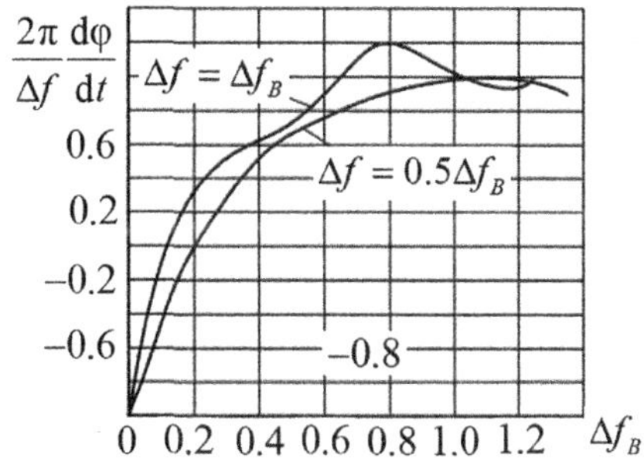

Fig. 4.6 The graphs of the dependence of the envelope settling at the output of a single-stage resonant amplifier on various values of the frequency jump at its input, built on the basis of expression (4.11)

difference frequency $\Delta\omega$, tending to the value $\dot{U}_{B2}$. At the same time, the swing of the vector $\dot{U}_2$ relative to $\dot{U}_{B2}$ a gradually decreasing deflection angle indicates that the frequency of the output voltage is also established gradually over the lifetime of natural oscillations. Thus, the duration of the transient processes at a frequency jump and an amplitude jump is the same.

4.3 Effects of Interference on a Frequency Signal Receiver

Let us consider the influence of sinusoidal and fluctuation interference on the quality of the receiver of frequency-modulated signals [9].

Suppose that a useful frequency-modulated signal u_C and a harmonic interference u_B are simultaneously received at the input of the receiver, the difference of the carrier frequencies of which $f_C - f_B = F_B$ is, and the ratio of the amplitudes $\gamma = \frac{U_{mC}}{U_{mB}} > 2$. If $F_B < 0.5\Delta f_B$ the interference also enters the main reception channel, then the process of its interaction with the signal can be represented by the vector diagram shown in Fig. 4.8. Since $f_C \neq f_B$, the vector U_{mB} rotates relative to the vector U_{mC} with a difference frequency F_B, leading to a periodic change in the amplitude and phase of the resulting Fig. 4.7 shows oscillation, which turns out to be modulated in amplitude and frequency.

When $\gamma > 2$ parasitic amplitude modulation is easily eliminated by an amplitude limiter with a clipping threshold

$$U_{th} \leq U_{mC} - U_{mB}. \tag{4.12}$$

Thus, under the assumed initial assumptions, the effect of interference is practically determined only by parasitic frequency modulation. In accordance with the vector diagram, from the triangle formed by the vectors U_{mC}, U_{mB} and U_m, we have

$$\sin m_{f_B} = \frac{U_{mB}}{U_m} = \frac{1}{\gamma}. \tag{4.13}$$

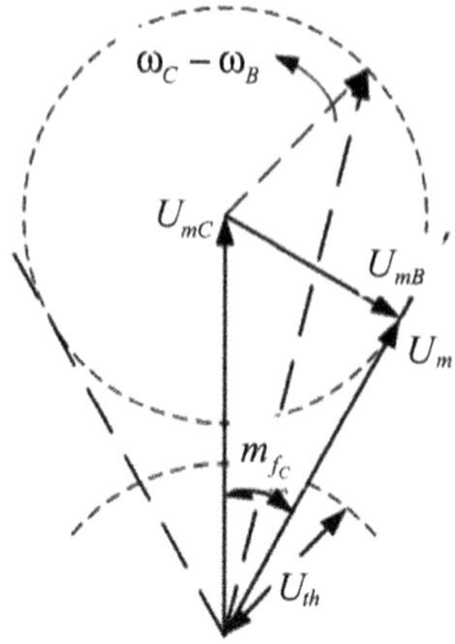

Fig. 4.7 Oscillation, which turns out to be modulated in amplitude and frequency

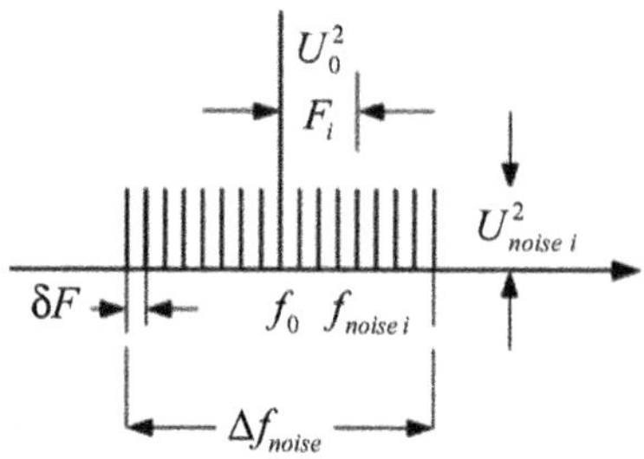

Fig. 4.8 The vector diagram

Hence (since by condition $(1/\gamma) < 0,5$), in the first approximation, we can assume that

$$m_{f_B} \approx \frac{1}{\gamma}. \tag{4.14}$$

According to the definition of the modulation index, the frequency deviation due to interference is equal to

$$\Delta f_{mB} = m_{f_B} F_B \approx \frac{F_B}{\gamma}. \tag{4.15}$$

At the output of the frequency detector, the amplitudes U_{mCout} of the interference U_{mCout} signal are proportional to the corresponding frequency deviations Δf_{mC} and Δf_{mB}. Therefore

$$\gamma_{out} = \frac{U_{m\ out}}{U_{mB\ out}} = \frac{\Delta f_m}{\Delta f_{mB}} \approx \frac{\Delta f_m}{F_B}\gamma. \tag{4.16}$$

Equation (4.16) shows that the signal-to-noise ratio at the output of the detector can be greater than the same ratio at its input, and the resulting gain q depends on the ratio of the useful signal frequency deviation to the detuning between the signal and noise frequencies:

$$q = \frac{\gamma_{out}}{\gamma} \approx \frac{\Delta f_m}{F_B}. \tag{4.17}$$

With increasing detuning F_B, the effect of the interference increases and the gain q decreases. If the detuning exceeds the maximum frequency of the modulating oscillation $F_{\max}$, then the interference arising at the output of the frequency detector will be filtered out by the low-frequency channel of the receiver, the upper cutoff frequency of which, as a rule, matches the range of the modulating frequencies. Based on these considerations, it is possible to estimate the smallest gain that will be provided in the worst conditions when $F_B = F_{\max}$:

$$q_{\min} = \frac{\Delta f_{mC}}{F_{\max}} = m_{fc}. \tag{4.18}$$

It should be remembered that the results obtained are valid for sinusoidal interference, the amplitude of which is much less than the useful signal ($\gamma > 2$), and the realization of a large gain requires the use of large modulation indices corresponding to wideband frequency modulation. Physically, this is explained by the fact that, theoretically, the frequency deviation of the useful signal with an increase in the modulation index m_{f_c} increases the infinitely ($\Delta f_{mC} = m_{fc} F_{\max}$) maximum frequency deviation caused by interference at $\gamma > 2$, in accordance with expression (4.14), is limited by the modulation index $m_{f_B} < 0.5$, which corresponds to $\Delta f_{mB} = 0.5 F_{\max}$.

Let us now turn to the consideration of the problem when the fluctuation voltage plays the role of the noise. Its rigorous solution requires the use of a rather complex mathematical apparatus, so we will use approximate estimates. To do this, we represent the resonance curve of the high-frequency channel of the receiver rectangular with a passband [10, 11]

$$\Delta f_B = \Delta f_{noise} \approx 2\Delta f_{mC}. \tag{4.19}$$

Equality (4.19) according to (4.3), in the first approximation, is valid for broadband frequency modulation, when $m_{f_c} \gg 1$.

Let's divide the noise bandwidth into n arbitrarily small sections δF. In this case, the noise voltage U^2_{noise} equivalent to the entire noise band Δf_{noise} can be represented as the sum of n sinusoidal voltages

$$U^2_{noise\,i} = \frac{U^2_{noise}}{n} = U^2_{noise} \frac{\delta F}{\Delta f_{noise}}. \tag{4.20}$$

Moreover, the phases of these voltages are random, and the frequencies of the adjacent components differ from each other by δF.

Estimating the level of noise at the output of the detector with a large amplitude of the harmonic signal ($\gamma > 2$), we can neglect the components that appear due to the detection of beats of the noise-noise type and restrict ourselves to taking into account only the components from the beats of the signal-to-noise type [12].

Suppose that the frequency of the i-th component of the noise differs from the signal frequency by an amount F_i (Fig. 4.8).

Then, according to formulas (4.13), (4.14) and (4.20), the square of the modulation index of the signal—the i-th component of the interference—will be

$$m^2_{f_{noise\,i}} = \frac{U^2_{noise\,i}}{U^2_C} = \frac{U^2_{noise}}{U^2_C}\frac{\delta F}{\Delta f_{noise}}. \quad (4.21)$$

Accordingly, the square of the frequency deviation due to the interference

$$\Delta f^2_{m\;noise\,i} = m^2_{f_{noise\,i}} F_i^2 = \frac{U^2_{noise}}{U^2_C}\frac{F_i^2}{\Delta f_{noise}}\delta F. \quad (4.22)$$

Of all the components of the detector's output voltage, only those whose frequencies fall within the noise bandwidth of its low-frequency path will pass to the receiver output ΔF_{noise}. Therefore, the square of the resulting frequency deviation due to noise is defined as the sum of the squares of the elementary deviations caused by individual noise components, the detuning of which is relative to the carrier frequency of the signal $F_i \leq \Delta F_{noise}$. If we consider the number of harmonic noise components n to be large enough, then the indicated summation can be performed by taking the following integral [13–15]:

$$\Delta f^2_{m\;noise} = 2\int_0^{\Delta F_{noise}} \frac{U^2_{noise}}{U^2_C}\frac{F_i^2}{\Delta f_{noise}} dF_i = \frac{2}{3}\frac{U^2_{noise}}{U^2_C}\frac{F_i^3}{\Delta f_{noise}}. \quad (4.23)$$

Hence, taking into account equality (4.19), we have

$$\Delta f^2_{m\,noise} = \frac{U_{noise}}{U_C}\sqrt{\frac{\Delta F^3_{noise}}{3\Delta f_{mC}}}. \quad (4.24)$$

Formula (4.24) is derived for the case of a modulated signal. However, experimental data show that at $\gamma > 2$ it is also valid for a frequency-modulated signal. This is due to the fact that with frequency modulation in the band Δf_{noise} there are many components of the signal spectrum, but the sum of the powers of these components is practically equal to the power of the unmodulated signal.

Since the voltage at the output of the frequency detector is proportional to the corresponding frequency deviation, then [6, 16]

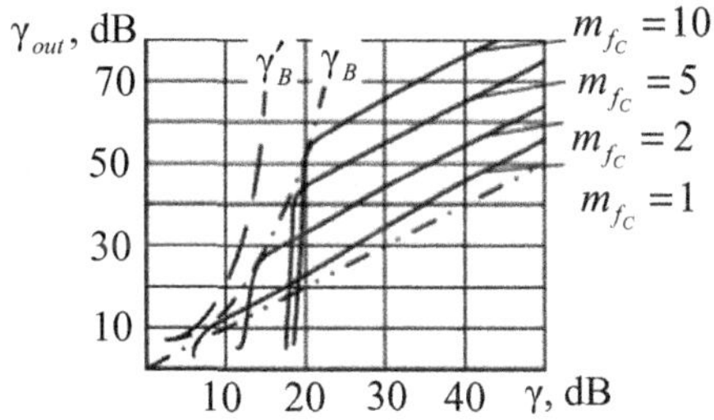

Fig. 4.9 The graphs of the dependence γ_{BbIX} on at various modulation indices, constructed for large values γ based on relation (4.26)

$$\gamma_{out} = \frac{U_{C\,out}}{U_{noise\,out}} = \frac{\Delta f_{mC}}{\Delta f_{m\,noise}} = \frac{U_C}{U_{noise}}\sqrt{\frac{3\Delta f^3_{mC}}{\Delta F^3_{noise}}}. \tag{4.25}$$

Assuming in the obtained ratio the value of the noise bandwidth of the low-frequency path is approximately equal to the upper modulation frequency $\Delta F_{noise} \approx F_{\max}$ and taking into account that $\Delta f_{mC}/F_{\max} = m_{f_c}$, we have

$$\gamma_{out} = \gamma\sqrt{3m^3_{fC}}. \tag{4.26}$$

From here

$$q = \frac{\gamma_{out}}{\gamma} \approx \sqrt{3m^3_{fC}}. \tag{4.27}$$

Thus, even with noise interference, a significant gain is obtained only with broadband frequency modulation, when the modulation index of the useful signal $m_{f_c} > 1$.

Figure 4.9 shows the graphs of the dependence γ_{BbIX} on at various modulation indices, constructed for large values γ based on relation (4.26).

It can be seen from the figure that with weak interference corresponding to large values γ, the value of the signal-to-noise ratio at the output of the FM signal receiver increases with respect to the same value of the ratio at its input, and this increase is the greater, the greater the frequency modulation index [17]. However, in the case of large interference, corresponding to small values γ, the value of the signal-to-noise ratio at the output drops sharply. The latter is explained by the fact that at $\gamma \leq 1$ the limiter at the actually selected limit threshold [see. expression (4.12)] cannot completely eliminate the parasitic amplitude modulation. Naturally, this affects the noise immunity of receiving FM signals. In addition, with strong interference, the phase deviation angle of the resulting voltage vector is mainly determined by the interference. The modulation index due to interference becomes much greater than unity, which leads to a sharp increase in the noise voltage at the detector output.

It should be noted that at low modulation indices of the useful signal, the gain, although smaller in absolute value, remains at a significantly higher level of interference than in the case of large modulation indices γ_B. In other words, the threshold

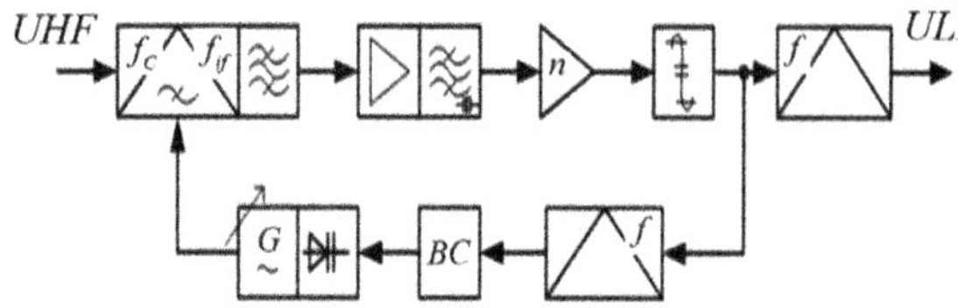

Fig. 4.10 The block diagram of a negative frequency feedback

value at which the frequency-modulated information system gives a benefit is lower. The reason for this is the expansion of the bandwidth required to receive high signal values m_f. An increase in the receiver bandwidth leads to the appearance of a significant number of noise spikes at the input of the frequency detector, significantly exceeding the rms value of the noise voltage, commensurate with the signal amplitude, and sometimes even exceeding it. As a result, the value γ_{out} drops sharply even when the input signal power exceeds the average noise power. So, from Fig. 4.10 it can be seen that at $m_{f_c} = 1$, the gain remains up to $\gamma = 5$ dB values, while at $m_{f_c} = 5$ it begins to rapidly decrease already at $\gamma = 18$ dB. This explains the fairly widespread use of systems with a relatively narrowband frequency modulation in military equipment, since a high level of interference is the usual condition for their operation [9, 10].

4.4 Methods of Increasing the Immunity of Reception of Frequency-Modulated Signals

The sensitivity of a FM receiver is largely determined by the signal-to-noise ratio at its input. Such receivers have a clearly expressed threshold property: when the signal amplitude decreases to values corresponding to a certain value, a sharp increase in the level of interference at the receiver output occurs, leading to significant distortions of the received information.

The amplitude of the fluctuation noise is a random value. It can have instantaneous values significantly higher than the root-mean-square value. However, the probability of the interference exceeding a certain level decreases rapidly as the degree of excess increases. Therefore, a decrease in the rms noise voltage relative to the signal voltage leads to a significant weakening of the interference effect, since the intervals during which the interference amplitude exceeds the signal amplitude become more rare and short-lived.

The signal level, which corresponds to the maximum permissible probability of distortion of the received message by noise, is often called the threshold level [18].

Reducing the threshold signal level due to the use of special circuit solutions allows you to provide the required reception quality at a lower signal level. This leads, at the same communication range, to a decrease in the required transmitter power, an increase in secrecy and, consequently, in the noise immunity of the radio link. One of the ways to lower the threshold level of the signal can be the introduction of a frequency-modulated signal into the receiving path of a negative frequency feedback (FF), the block diagram of which is shown in Fig. 4.10.

In essence, this feedback is equivalent to the frequency locked loop (FLL) system and differs from it only in the absence of a low-pass filter between the frequency discriminator and the correction block (BC), which determines the required character of the transmission coefficient of the control circuit, the time constant of which is chosen less than the period of the lowest modulation frequency. Frequency feedback with a large frequency auto-tuning coefficient leads to a significant decrease in the frequency deviation, and hence the frequency modulation index, thereby narrowing the spectrum width of the received signal after the frequency converter (4.3). This allows the bandwidth of the intermediate frequency path to be narrowed to the same extent. The reduction of frequency deviations by the FF circuit does not depend on the reasons for the frequency deviation, it occurs equally with a useful frequency change caused by transmitter modulation, and when it changes under the influence of interference [19]. It would seem that frequency feedback does not affect signal distortion by interference.

However, the likelihood of signal suppression when narrowing the bandwidth of the intermediate frequency path decreases, since the signal amplitude remains unchanged, and the rms noise voltage, which is proportional to the square root of the noise bandwidth, decreases [12, 14]. As a result, the signal-to-noise ratio at the input of the frequency detector increases, the intervals during which the instantaneous value of the noise exceeds the threshold become more rare and short-lived, which reduces the probability of errors. Naturally, for a given failure probability, this is equivalent to lowering the threshold. A similar role in the reception of frequency-modulated signals is played by a phase-synchronous detector, the block diagram of which is shown in Fig. 4.11. This circuit is essentially a layered filter based on a phase-locked loop.

The frequency-modulated signal from the output of the receiver's linear path is fed to the phase discriminator, to the second input of which the voltage of the controlled local oscillator is applied. The control circuit provides tracking of the local oscillator frequency to the frequency of the useful signal. With an appropriate choice of the structure and parameters of the control circuit, the frequency changes turn into proportional changes in the phase difference between the signal and local oscillator voltages. Accordingly, the detected signal is sent at the output of the phase detector. A similar result can be obtained after detecting the heterodyne voltage with a conventional frequency detector [20].

Here, as in the previous circuit, there is deep negative frequency feedback. The only difference is that in a phase-locked detector, the spectrum is transferred not to the intermediate frequency region, but directly to the modulating frequency region.

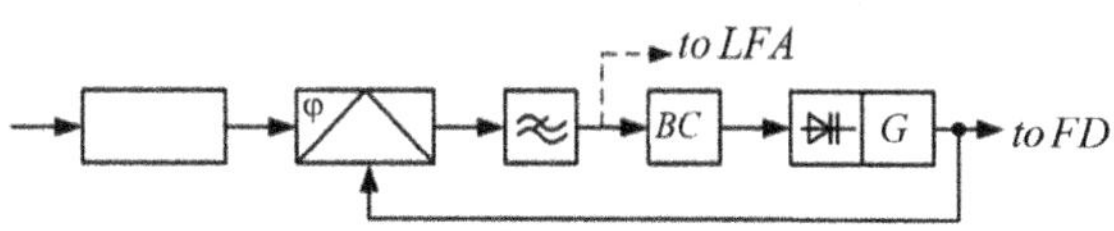

Fig. 4.11 The block diagram of a phase-synchronous detector

In this case, the required bandwidth of the closed control loop is determined by the maximum modulating frequency, and its noise bandwidth is much less than the noise bandwidth of the receiver's linear path. The latter leads to a significant decrease in the threshold for the same error probability.

Phase-synchronous detector circuits are successfully used for noise-immune reception of frequency-modulated signals with small modulation indices. At higher values, better results are obtained with negative frequency feedback circuits.

The threshold gain in demodulators of the considered type is determined by the degree of compression of the signal spectrum. In the limiting case, the spectrum width decreases to a value approximately equal to $2F_{\max}$. The actual bandwidth of the filter of the tracking demodulator, according to expression (4.3), decreases by a factor $m_{f_c} + 1$, which leads to a decrease in the threshold by $20 \lg \sqrt{m_{f_c} + 1}$ dB. For a number of reasons, this limiting decrease in the threshold is practically not achieved. In systems with high-frequency frequencies, for example, this is prevented by the action of noise entering the mixer input through the feedback loop. Really feasible approximate threshold values when using the tracking demodulators of frequency-modulated signals γ_B' discussed above are marked in Fig. 4.10 with dotted curve.

4.5 Doppler Signal Radio Receivers

Doppler meters are widely used in modern command and measurement systems. As you know, the frequency of the signal at the point of reception and the frequency of the transmitter located on the moving object differ by the magnitude of the Doppler frequency shift. This displacement in the first approximation is proportional to the carrier frequency and the radial velocity of the moving object relative to the receiving point. The Doppler effect manifests itself both at the frequency of the signal and at the frequencies of any other oscillation that can be extracted from the signal using linear or nonlinear transformations. The value of the Doppler frequency shift and the law of its change in time contain useful information about the radial velocity and trajectory of the object [21].

Doppler meters typically operate at decimeter and centimeter wavelengths. At the same time, to improve the accuracy, the Doppler shift measurements are made at sufficiently low intermediate frequencies. The latter circumstance predetermines the need to use a superheterodyne receiver circuit, in which multiple frequency conversion is used to ensure good selectivity in the main and side reception channels. In practice, already in systems with double conversion of frequency, using modern lumped selection filters, it is possible to fulfill very stringent requirements for selectivity. However, sometimes, especially in interrogation systems with signal re-emission on board, the logic of constructing a measuring complex may require three to four times the frequency conversion.

Essentially, a Doppler signal is a frequency-modulated waveform with a very low modulation frequency. Therefore, the width of its spectrum at the time of reception is very small. In practice, a sinusoidal signal arrives at the input of the receiver, the frequency of which can slowly vary over time over a very wide range. The optimal procedure for receiving such a signal with a priori unknown frequency and phase assumes the use of a tracking filter built on the basis of a phase-locked loop with an astatism order and a noise bandwidth determined by the permissible measurement errors and the possible law of frequency change at the input. To expand the acquisition bandwidth and accelerate the entry into communication in the receiver, one of the variants of the digital frequency control system, various variants of the search scheme, can be used.

In order to combat impulse interference, the main receiving channel, as a rule, contains a broadband path, an amplitude limiter and a narrowband amplifier (BPALNA) circuit, which is a serial connection of a broadband path, an amplitude limiter and a narrowband amplifier, the bandwidth of which is matched to the bandwidth of the received narrowband signal [22]. As the transient analysis in Chap. 5 shows, the broadband input path, when exposed to powerful video pulses, produces a response that increases in amplitude and decreases in duration with increasing bandwidth. Thus, the broadband path, as it were, concentrates the interference energy in time, turning it into a series of short radio pulses of high intensity. In addition, it provides signal amplification and stabilization of its level at the input of the limiter due to the action of an effective AGC system. The amplitude limiter sharply reduces the effectiveness of the interference due to the choice of the limiting threshold slightly higher than the level of the useful signal. A narrowband amplifier, amplifying the signal practically without distortion, deforms a short limited radio interference pulse, which at the output does not have time to grow to its steady-state value. The principle of operation of the BPALNA system is illustrated by the timing diagrams shown in Fig. 4.12, where, for simplicity, the signal envelopes and interference are shown at various points of the BPALNA circuit. Note that the BPALNA system is also effective when the receiver is exposed to interference in the form of short radio pulses, since the broadband path reproduces such interference well and amplifies it. Heterodyne voltages for frequency converters are supplied from a heterodyne frequency synthesizer (HFS), the stability of which is determined by a single reference frequency of the system. To get into communication quickly, a search system is provided in the receiver, which includes n filters with adjacent passbands, a signal filter and a search circuit acting on the frequency of the second local oscillator F_2.

As a result of its operation, the useful signal falls into the passband of the main signal filter. The receiver contains an FLL system, a tracking filter based on a phase-locked loop (PLL) system and a Doppler frequency meter, which receives information from the search systems, AFC, a tracking filter and forecast data that determines the initial choice of the frequency of the first local oscillator.

4.6 Main Features of the Following Filter

The sensitivity of a radio receiving device substantially depends on its bandwidth, the optimal value of which provides the highest signal-to-noise ratio at the output

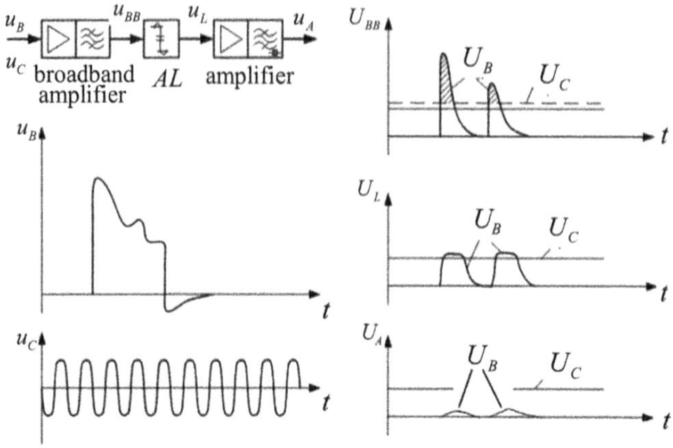

Fig. 4.12 The principle of operation of the BPALNA system

of the receiver's linear path. When evaluating the optimal bandwidth, in most cases, one proceeds from the integral, steady-state spectrum of the received signal. In a continuous Doppler system, despite the fact that the range of changes in the Doppler frequency, and therefore the integral spectrum of the signal, is quite wide, the instantaneous (current) spectrum practically consists of one sinusoidal component, the frequency of which changes very slowly over time. The reception of such signals can be carried out using narrowband, automatically tunable selective tracking filter systems [7, 23]. The use of tracking filters in Doppler signal receivers can significantly increase their sensitivity and provide high accuracy in measuring the radial velocity of an object.

In addition to Doppler systems, tracking filters are used to isolate individual components of the spectrum of the received signal, for example, the carrier frequency. The voltage obtained in this way makes it possible to form a reference signal for phase or amplitude synchronous detectors. The given examples by no means exhaust the possible areas of application of tracking filters in various radio engineering devices.

The basis for constructing the overwhelming majority of tracking filter circuits is a phase-locked loop system. The block diagram of a typical tracking filter is shown in Fig. 4.13. Since the transmission coefficient of this link is $K_1(p) = 1$, its absence in the tracking filter will not change the transmission coefficient of the closed control loop, which, when used in the system of a proportional-integrating filter, is equal to

$$K_Z(p) = \frac{\Delta f_2(p)}{\Delta f_1(p)} = \frac{T_2 p^2 + p}{T_2 p^2 + (1 + K_0 T_1)p + K_0}.$$

Recall that this ratio was obtained under the assumption that all elements of the control circuit are inertia, with the exception of the filter. The value Δf_1 determines the change in the signal frequency relative to its nominal value and is a disturbing effect; Δf_G, the change in the frequency of the controlled generator due to the

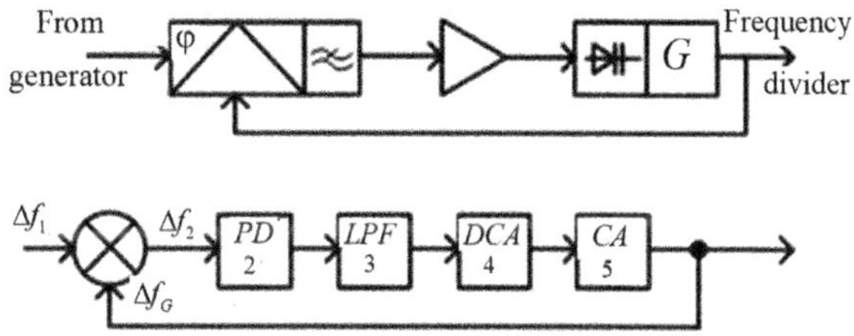

Fig. 4.13 The block diagram of a typical tracking filter

influence of the control voltage of the AFC system and Δf_2 is the error of the automatic control system, equal to the difference between the frequencies of the input signal and the controlled generator:

$$\Delta f_2(t) = \Delta f_1(t) - \Delta f_G(t). \tag{4.28}$$

Since the main task of the tracking filter is the most accurate correspondence of the frequency of the controlled generator to the frequency of the input signal, we introduce the concept of the transfer function of the tracking filter (the transfer coefficient of the closed-loop control over the local oscillator):

$$K_{MF}(t) = \frac{\Delta f_G(t)}{\Delta f_1(t)} = 1 - \frac{\Delta f_2(t)}{\Delta f_1(t)} = 1 - K_Z(t).$$

Hence, taking into account formula (4.24) in operator form, we have

$$K_{MF}(t) = 1 - K_Z(p) = \frac{K_0(1 + j\Omega T_1)}{T_2 p^2 + (1 + K_0 T_1)p + K_0}. \tag{4.29}$$

Replacing the operator p by $j\Omega$ in expression (4.29), we obtain the frequency response of the tracking filter:

$$K_{MF}(t) = \frac{K_0(1 + j\Omega T_1)}{K_0 - \Omega^2 T_2 + j\Omega(1 + K_0 T_1)}. \tag{4.30}$$

Considering the tracking filter as some selective four-port network, it should be borne in mind that its input is affected by the frequency of the input signal, and the output effect is the frequency of the controlled oscillator. In this case, such an energy level of the signal is assumed at which the normal operation of the tracking filter is ensured, and the permissible fluctuations of the input voltage amplitude do not affect the frequency of the controlled generator. Therefore, selectivity here should be understood as the ability to select frequencies that determine the law of changing the frequency of the input signal [6, 9, 13].

An additive mixture of signal and noise is fed to the input of the tracking filter from the output of the receiver's linear path. The parameters of the control system that determine its speed should be selected in such a way that the controlled local oscillator tracks slow informative changes in the signal frequency as accurately as possible and reacts little to relatively fast frequency changes caused by the effect of noise. When these requirements are met, the voltage of the tunable local oscillator, reproducing changes in the frequency of the useful signal, has a significantly lower frequency deviation due to noise, and, therefore, a narrower spectrum than the signal–noise mixture entering the tracking filter. Thus, the output oscillations of the controlled local oscillator can formally be considered as a result of filtering the useful signal from its mixture with noise by the PLL system.

In other words, a tracking filter with a finite cutoff band is inertial with respect to the change in the signal frequency at its input. The latter results in a dynamic error. Its value depends on the nature of the disturbing effect and the speed of the transient process in the closed control loop [24].

The presence of noise voltage causes a noise error, which can be estimated by the variance of the voltage phase σ^2 of the controlled generator [19]:

$$\sigma^2 = \frac{1}{2\pi}\int_0^\infty S(\Omega)|K_{MF}(j\Omega)|^2 d\Omega.$$

Assuming $S(\Omega) = S_0$ and taking into account that $K_{C\Phi}(0) = 1$ in accordance with expression, we have

$$\sigma^2 = \frac{S_0 \Delta\Omega_{BB}}{2\pi}. \tag{4.31}$$

Here S_0 is the physical real spectral density of the noise voltage at the input of the tracking filter and $\Delta\Omega_{BB}$—noise bandwidth of the tracking filter.

To determine the noise bandwidth, we use relations (2.5) and (4.30):

$$\Delta\Omega_{BB} = \int_0^\infty |K_{MF}(j\Omega)|^2 \mathrm{d}\Omega = \int \frac{K_0^2(1+\Omega^2 T_1^2)}{(K_0-\Omega^2 T_2)^2 + \Omega^2(1+K_0 T_1)^2}\mathrm{d}\Omega.$$

After integrating this expression, we get

$$\Delta\Omega_{BB} = \frac{\pi}{2}\frac{K_0(T_2 + K_0 T_1^2)}{T_2(K_0 T + 1)}. \tag{4.32}$$

Taking $T_1 = lT_2$, where $l \in (0, 1)$ and using formulas (4.31) and (4.32), we determine the noise error of the tracking filter:

$$\sigma^2 = \frac{S_0 K_0}{4}\frac{1 + K_0 l^2 T_2}{1 + K_0 l T_2}. \tag{4.33}$$

Analysis of the relationship (4.33) shows that for a given control dynamics (dynamic error), which for systems with a damping coefficient is determined $\xi = 1$ by the values K_0 and T_2, the noise error has a minimum at some optimal value of the coefficient l. The corresponding value l_{opt} is determined $\sigma^2_{\min}$ from the equation:

$$\frac{d\sigma^2}{dl} = \frac{S_0 K_0}{4} \frac{2lK_0T_2(1+K_0lT_2) - K_0T_2(1+K_0l^2T_2)}{(1+K_0lT_1)^2} = 0.$$

Having solved this equation for l, we find

$$l_{opt} = \frac{-1 \pm \sqrt{1+K_0T_2}}{K_0T_2}.$$

Since l_{opt} it can take values from zero to one, and the product K_0T_2 in real systems is much greater than one, then

$$l_{opt} = \frac{-1 \pm \sqrt{1+K_0T_2}}{K_0T_2} \approx \frac{1}{\sqrt{K_0T_2}}. \tag{4.34}$$

Substituting the value in relation (4.33), we find the minimum value of the noise error:

$$\sigma^2_{\min} = \frac{S_0K_0}{2} \frac{1}{1+\sqrt{K_0T_2}} \approx \frac{S_0K_0}{2\sqrt{K_0T_2}}. \tag{4.35}$$

Figure 4.14 shows graphs of dependence $4\sigma^2/S_0K_0 = \varphi(l)$ for various values K_0T_2, built on the basis of expression (4.33).

Their analysis shows that the noise error with a rational choice of the filter time constants T_{1opt} and T_2, corresponding l_{opt} to the given dynamics of the control system (K_0T_2 = const), has a pronounced minimum, which is the deeper, the greater the value K_0T_2.

The obtained value l_{opt} is approximately two times less than the value l corresponding to the damping coefficient $\xi = 1$. This suggests that the transient process in the tracking filter becomes oscillatory at $T_{1opt} = l_{opt}T_2$. Accordingly, overshoot

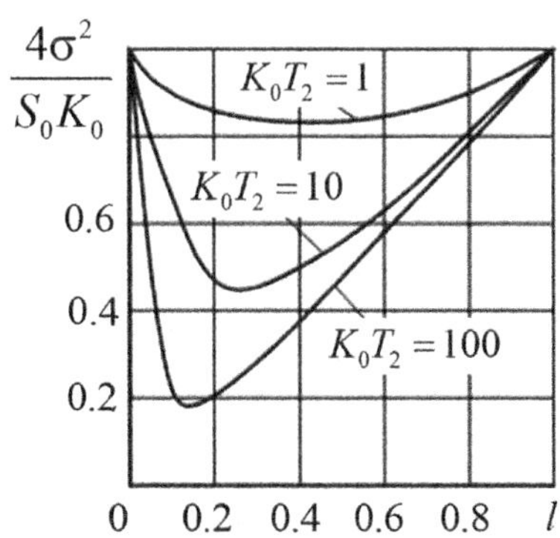

Fig. 4.14 The graphs of dependence $4\sigma^2/S_0K_0 = \varphi(l)$ for various values K_0T_2, built on the basis of expression (4.33)

increases, reaching 30%. However, this amount of overshoot can be considered acceptable, and the quality of the transient process is satisfactory.

The analysis shows that the use of a proportional-integrating link in a control loop with one integrator makes it possible to optimize the characteristics of the tracking filter, ensuring that the noise error is minimized with a dynamic error not exceeding a given value [15].

In the particular case when an aperiodic link is used in the control system, which corresponds to the values $R_1 = 0$, $T_1 = 0$, $l = 0$, according to relation (4.33), the fluctuation error does not depend on the filter time constant T_2 and is equal to

$$\sigma^2 = \frac{K_0 S_0}{4}. \tag{4.36}$$

Equality (4.36) remains valid and at $T_2 = 0$, which corresponds to an "ideal" PLL. This seemingly paradoxical result is explained by the absence of dependence of the noise bandwidth of the tracking filter on the parameters of the aperiodic link. According to expression (4.32) at $T_1 = 0$T_1 = 0

$$\Delta\Omega_{BB} = \frac{\pi}{2} K_0.$$

At the same time, the processes in the system become slower, which leads to an increase in the dynamic error. At the same time, a rise appears in the frequency response of the filter, the magnitude of which increases with increasing T_2. As a result, the area under the curve $|K(j\Omega)|^2$, and therefore remain constant. This significantly degrades the parameters of the tracking filter, since with the required control dynamics according to equality (4.36) and the curves shown in Fig. 4.14, the noise error of the tracking filter based on an ideal PLL or PLL with an aperiodic link will be significantly greater than the noise error of the optimal tracking filter with proportional-integrating link. Moreover, the loss will be the greater, the larger the work $K_0 T_2$.

4.7 General Information on Reception of Phasomanipulated Signals

In most modern communication systems, command and measurement radio lines, navigation and telemetry systems, signals with pulse code modulation of the pulse code modulation-phase modulation PCM-PM type are widely used [25]. The use of such signals significantly increases the noise immunity of systems, increases the reliability of information reception at large radio communication ranges and severe restrictions on the dimensions, weight and power consumption of equipment. To ensure that such a signal, the energy level of which at the receiving point is usually very small, is not suppressed by interference, it is necessary to optimize the structure of the receiving path. For this purpose, in the process of receiving a signal, a likelihood function should be formed $Z(J, \omega, \tau)$, which depends on the code combination number J, frequency ω, time delay τ, and its maximum should be found in three-dimensional space. Attempts to rigorously implement such a search lead to an excessive complication of the receiving device and require a rather large, and, in some cases, practically unrealistic time.

Rejecting the strictly optimal design of the receiving path in order to maintain the given probability of receiving error, it is necessary to increase the signal energy. However, the gains in the simplicity of construction and the reliability of the system may well offset these costs.

There are various ways to simplify the hardware. One of them, widely used in practice, is that the search for unknown signal parameters is carried out step by step, sequentially. First, the frequency and phase of the oscillation of the carrier of the received signal are determined. This information is used to create a coherent voltage reference. The reference signal formed in the receiver ensures the operation of the phase (synchronous) detector, with the help of which the useful signal is demodulated [26]. The phase detector performs the operation of multiplying the received signal and noise realization by the reference voltage. As a result, when the frequencies coincide and the initial phase shift of the reference voltage with respect to the carrier, a sequence of phase-shift keying symbols is allocated at $\pi/2$ the output of the detector. Then the identification of the code message is made.

Since the phase detector is a linear parametric system, it does not suppress a weak signal with noise. A phase detector with a reference voltage ideally coherent with the carrier does not introduce additional energy losses in comparison with the optimal reception scheme. Therefore, when receiving phase-shift keyed signals, one of the main tasks is to obtain a sufficiently good coherent reference voltage. The reference local oscillator, due to the instability of the oscillators, changes in the path parameters, and Doppler shifts of the signal frequency have to be adjusted to the carrier frequency using the PLL system.

Radio links that use coherent reference waves to demodulate a signal are called coherent radio links. A simplified block diagram of a receiver of such a radio link in the most general case has the form shown in Fig. 4.15. The main receiving channel (MRC), as a rule, ends with an amplitude limiter. The structural diagram of the MRC and its technical implementation determine the required receiver sensitivity, its selectivity for adjacent and side receiving channels, multi-signal selectivity, dynamic range and other characteristics, and also provide the required signal level at the input of the synchronous detector and the carrier synchronization channel.

Naturally, the characteristics and structure of the main receiving channel must be consistent with the parameters of the received signal, which in the general case has the following form:

$$u_C(t) = U_{mC} \sin[\omega_C f + \varphi_C + \varphi_m \Pi_C(t)], \tag{4.37}$$

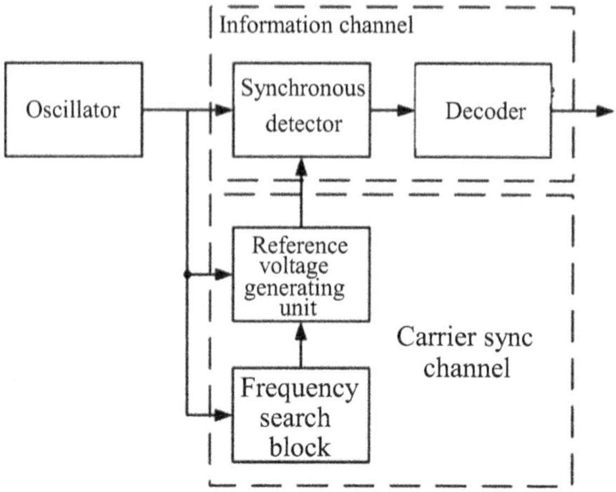

Fig. 4.15 A simplified block diagram of a receiver of a radio link

where φ_m is the phase deviation, which in various radio links can take values from zero to $\pi/2$ and $\Pi_C(t)$—a sequence of positive and negative pulses with a duration T_0 with a unit amplitude.

A positive pulse of the sequence $\Pi_C(t)$ corresponds to a one in the transmitted message and a negative pulse corresponds to a zero. The moments of the appearance of ones and zeros are random, and the probability of the appearance of one is $P(1)$, and of zero $-P(0) = 1 - P(1)$. The spectrum of such a signal is mixed. The discrete part of the spectrum consists of one harmonic component with frequency ω_0, amplitude U_{ω_0} and phase φ_{ω_0}, and (4.26)

$$U_{\omega_0} = U_{\omega_c}\sqrt{\cos^2\varphi_m + [2P(1) - 1]^2 \sin^2\varphi_{\mathrm{m}}}, \tag{4.38}$$

$$\varphi_{\omega_0} = -arctg\{[1 - 2P(1)]tg\varphi_m\}. \tag{4.39}$$

The continuous part of the spectrum has a spectral density

$$S(\omega) = 2U_{mC}^2 P(1)[1 - P(1)]\sin^2\varphi_m\tau_0\left(\frac{\sin\frac{\omega-\omega_0}{2}\tau_0}{\frac{\omega-\omega_0}{2}\tau_0}\right)^2. \tag{4.40}$$

With a symmetric stream of symbols in the transmitted message, when the probability of occurrence of ones and zeros is the same and equal to 0.5, changing the phase φ_m deviation from zero to $\pi/2$, any power distribution between the two parts of the spectrum is obtained.

So, in the case π of manipulation, which is widely used in practice, when $2\varphi_m = \pi$, according to relations (4.38) and (4.40), the carrier is absent ($U_{\omega_0} = 0$) in the transmitted signal, but $S(\omega)$ is maximum. In the absence of manipulation, when $\varphi_m = 0$, $U_{m_0} = U_{mc}$ and $S(\omega) = 0$.

4.8 Distortions of Phasomanipulated Signals in the Linear Tract of the Receiver

Consider the physical processes occurring in a selective amplifier when a phase-shift keyed signal is applied to its input. As shown in Fig. 4.16, a harmonic voltage that undergoes a phase jump at a certain angle $2\varphi_m$ at the moment $t = 0$ can be represented in the form of two oscillations [15, 27]:

$$u_1 = \begin{cases} u_{11} = U_m e^{j(\omega_0 t - \varphi_m)} \text{ at } t < 0, \\ u_{12} = U_m e^{j(\omega_0 t + \varphi_m)} \text{ at } t \geq 0. \end{cases} \tag{4.41}$$

We will assume that the first of these voltages was applied to the input of the amplifier for a long time, so that by the time $t = 0$ the transient process caused by the switching on of this voltage is practically over. Under these conditions, at $t \geq 0$, two voltages will act at the amplifier output: due to the end of the voltage u_{21}, with the phase and amplitude decreasing in time according to the law $U_m[1 - B(t)]$, and u_{22} due to the appearance of a voltage u_1 with the phase $-\varphi_m$ and amplitude increasing in time according to the law $U_m B(t)$. Recall that $B(t)$ is the transient response of the amplifier amplitude. The corresponding timing diagrams of these voltages are shown in Fig. 4.17.

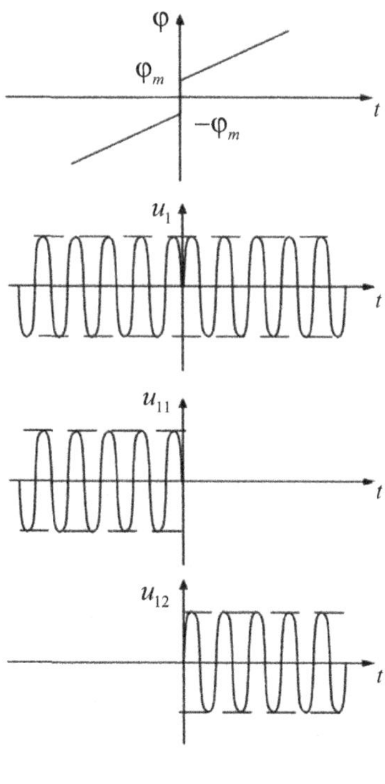

Fig. 4.16 A harmonic voltage that undergoes a phase jump at a certain angle $2\varphi_m$ at the moment $t = 0$

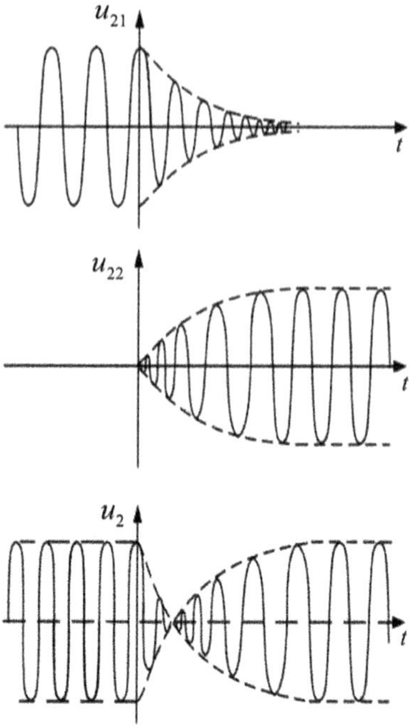

Fig. 4.17 The corresponding timing diagrams of these voltages

Thus, with $t \geq 0$

$$\left.\begin{aligned} u_{21} &= U_m[1 - B(t)]e^{j(\omega_0 t - \varphi_m)}, \\ u_{22} &= U_m B(t) e^{j(\omega_0 t + \varphi_m)}. \end{aligned}\right\} \tag{4.42}$$

Based on the principle of superposition, the output voltage of the amplifier for times $t \geq 0$ will be

$$u_2 = u_{21} + u_{22} = U_m e^{j\omega_0 t}\left\{[1 - B(t)]e^{-j\varphi_m} + B(t)e^{j\varphi_m}\right\}. \tag{4.43}$$

In the resulting relationship, the expression in curly brackets is the normalized complex amplitude of the amplifier output voltage

$$\dot{U}_2(t \geq 0) = [1 - B(t)]e^{-j\varphi_m} + B(t)e^{j\varphi_m}. \tag{4.44}$$

Replacing in accordance with the Euler $e^{\pm j\varphi_m}$ formula by $\cos\varphi_m \pm j\sin\varphi_m$, after elementary transformations, we obtain

$$\dot{U}_2(t \geq 0) = \cos\varphi_m - j[1 - 2B(t)]\sin\varphi_m. \tag{4.45}$$

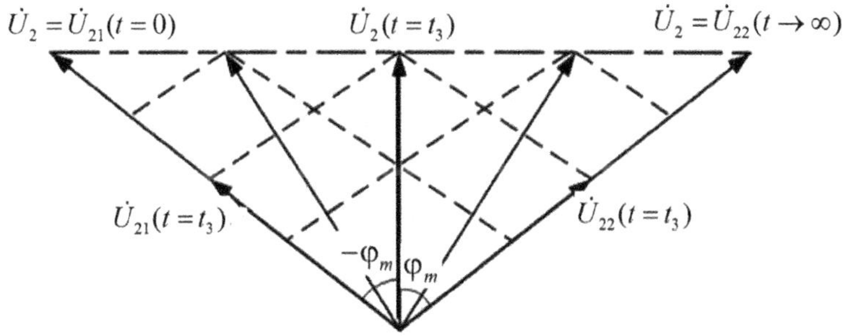

Fig. 4.18 The vector diagrams of the nature of the transient process

Expression (4.45) makes it possible to obtain the dependence of the envelope setting and the phase of the output voltage on time:

$$U_{m2}(t \geq 0) = \sqrt{\left[\mathrm{Re}\dot{U}_2(t)\right]^2 + \left[\mathrm{Im}\dot{U}_2(t)\right]^2} = \sqrt{\cos^2 \varphi_m + [1 - 2B(t)]^2 \sin^2 \varphi_m}. \tag{4.46}$$

$$\mathrm{tg}[\varphi(t \geq 0)] = \frac{\mathrm{Im}\left[\dot{U}_2(t)\right]}{\mathrm{Re}\left[\dot{U}_2(t)\right]} = -[1 - 2B(t)]\mathrm{tg}(\varphi_\mathrm{m}). \tag{4.47}$$

The nature of the transient process is illustrated by the vector diagrams shown in Fig. 4.18, which shows the process of geometric summation of the components of the output voltage u_{21} and u_{22} determined by equality (4.42).

Analysis of expression (4.46) shows that with a phase jump, both the phase and the amplitude of the output voltage undergo a change.

Since $B(0) = 0$ and $B(\infty) = 1$, the amplitude of the output voltage before the jump and after the end of the transient is equal to its steady-state value $U_{m2}(0) = U_{m2}(\infty) = 1$. The minimum value of the normalized amplitude of the output voltage is $-U_{m2\,\min} = \cos\varphi_m$. It occurs with a delay relative to the phase jump of the input voltage for a time interval equal to the group delay time in the amplifier, when $B(t) = 0,5$. Thus, with a phase jump by an angle $2\varphi_m = \pi$, the amplitude of the output voltage drops to zero during the transient process. The graphs of the dependences of the voltage setting at the output of the resonant amplifier, built on the basis of expression (4.46), are shown in Fig. 4.19.

Consider Eq. (4.47), which describes the process of establishing the phase of the output voltage $\varphi(t)$. It is obvious that at $t = 0$, and at $t \to \infty$ $\varphi(\infty) = \varphi_m$, i.e., the phase of the output voltage, as well as the phase of the input signal, during the transient process changes by $2\varphi_m$.

Moreover, at small φ_m, when $tg\varphi_m \approx \varphi_m$, the time to establish the phase of the output voltage is maximum.

With increasing φ_m, the phase settling time decreases, and when $\varphi_m = \frac{\pi}{2}$ the phase is set abruptly, and the phase jump occurs at $t = t_3$ the moment when the amplitude of the output voltage is zero.

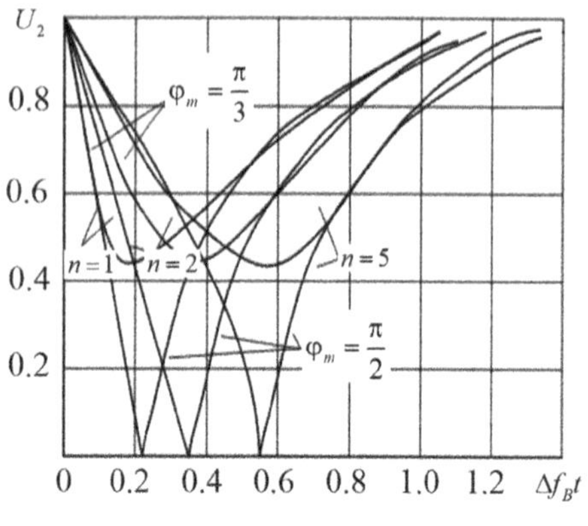

Fig. 4.19 The graphs of the dependences of the voltage setting at the output of the resonant amplifier, built on the basis of expression (4.46)

The latter is clearly seen from the consideration of vector diagrams (see Fig. 4.18) for a particular case $2\varphi_m = \pi$.

In this case, the output voltage is the sum of two antiphase components, and its phase is determined by the phase of the largest of the terms.

4.9 Principles of Obtaining Coherent Reference Vibration in the Presence of a Carrier in the Received Signal

When receiving a signal containing a carrier frequency component (pilot signal) in the spectrum, tracking filters based on the PLL system are usually used to form the coherent reference voltage of the phase detector. The voltage at the input of such a system is supplied from the output of the linear part of the receiver through a narrowband filter called a pilot-signal filter (PSF). The narrower the bandwidth of this filter, the better the carrier frequency component is selected from the spectrum of the received signal. At the same time, an excessively narrow PSF bandwidth can lead to a breakdown in synchronization with possible deviations of the intermediate frequency from its nominal value, due to the instability of local oscillators, errors arising during the operation of automatic frequency control systems, deep signal fading, etc.

As discussed in Sect. 4.6, the performance of the tracking filter is characterized by phase error, which is estimated by the phase difference between the carrier (pilot signal) and the voltage of the reference local oscillator. The dynamic component of this error is determined by the inertia of the auto-tuning loop and the nature of the disturbing effect. Control systems of modern tracking filters have a second and sometimes even a third order of astatism, which allows tracking a linearly varying frequency without errors [28]. The fluctuation component of the phase error is estimated by the phase variance of the controlled local oscillator. In the general case [26] with phase fluctuations much less than $\pi/2$

$$\sigma^2 \approx \frac{S_0 \Delta F_{noise}}{P_{noise\,0}}, \tag{4.48}$$

where S_0 is the power spectral density of the broadband noise at the input of the PLL system; ΔF_{noise} is noise bandwidth of the tracking filter and $P_{noise\,0}$ is the power of the carrier (pilot) component used for synchronization.

The difference between relations (4.48) and (4.31) is due to the assumption of a large energy level of the signal at the input of the tracking system, made when deriving formula (4.31). At the same time, the permissible level changes practically did not affect the frequency of the controlled generator. This assumption was valid for Doppler systems, in which the signal energy is practically concentrated at the carrier frequency. When FMS is received, the pilot-signal energy in the total energy level is very low. Consequently, the signal-to-noise ratio at the input of the tracking filter is also small, which leads to an increase in the noise component of the phase error and its dependence on the signal level [29].

Naturally, increasing the power of the carrier frequency component decreases the fluctuation error. However, in this case, the fraction of the power falling on the components of the continuous spectrum of the signal used to distinguish the symbols of the transmitted message decreases. The latter, ultimately, can lead to an increase in the probability of errors during reception.

The nature of the spectrum of signals with modulation of the PCM-PM type leads to the occurrence of a specific interference in the synchronization channel. As can be seen from Fig. 4.20, in such a signal, the continuous and discrete spectra overlap. Consequently, at the input of the tracking filter, no matter how narrow the passband of the pilot-signal filter, along with the carrier, a part of the most intense components of the continuous spectrum also falls. In the first approximation, this is equivalent to the effect of some additional fluctuation interference, leading, in accordance with expression (4.48), to an increase in the phase fluctuations of the reference signal, which, of course, affects the quality of the receiver's phase detector.

Specific interference can be eliminated by using more sophisticated modulation, such as PCM-FM-PM. At the first stage, frequency-shift keying of the subcarrier is

Fig. 4.20 The Costas circuit

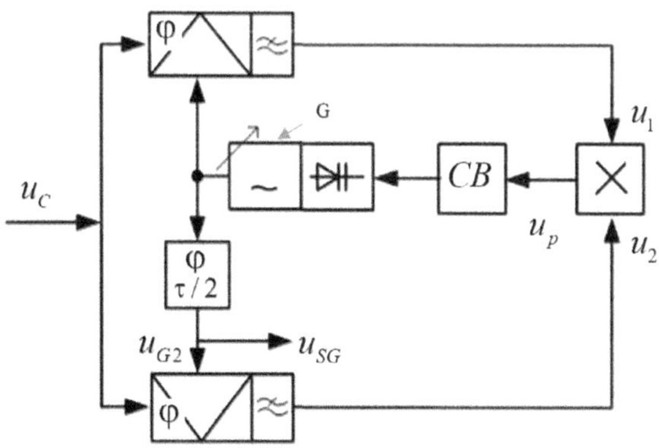

performed, at which the symbols of the PCM signal are filled with frequency ω_1 and ω_2 oscillations and for zeros and ones, respectively. Often rectangular waves—a square wave are used as a subcarrier, which facilitates the operation of the subsequent phase modulator (pulse code modulation-pulse frequency modulation-phase modulation (PCM-PFM-PM) signal). The FSK subcarrier is then used to phase modulate the carrier frequency [30].

The spectrum of the PCM-PFM-PM signal, as well as the PCM-PM signal, is mixed, but its continuous parts of the spectrum are grouped around the frequencies $\omega_0 \pm \omega_1$ and $\omega_0 \pm \omega_2$. Thus, with a sufficiently large value of the subcarrier frequency $\omega_B = 0.5(\omega_1 + \omega_2)$, there will be no additional interference in the synchronization channel. However, a subcarrier in a signal with the same chip duration significantly expands the bandwidth occupied by such a signal and degrades the reliability of distinguishing symbols in the traffic channel. As the analysis shows, such systems give an energy gain only at low data transmission rates (less than a thousand binary units per second), which is typical only for ultra-long-distance communications. At a relatively short range, when the transmission rate reaches millions of binary units per second, the use of modulation PCM-FM provides clear energy benefits [19, 25].

Note that there may be other components of the phase error associated with a specific hardware design. So, in digital systems, they are caused by time discreteness, signal quantization, not ideal discriminators and other reasons.

As practice shows, the total phase error of the system should not exceed 40°. In this case, its influence on the quality of the phase detector operation can be considered negligible.

4.10 Nesush Recovery Systems with Coherent Reception of Signals

As shown in Sect. 4.7, when using—keying, there is no carrier frequency in the spectrum of the KSh—PM signal, which allows all the signal energy to be used to distinguish symbols in the information channel. Transforming relation (4.37), the signal voltage can be represented as

$$u_C(t) = U_{mC}\left\{\sin(\omega_C t + \varphi_C)\cos\left[\varphi_m \Pi_{(}t)\right] + \cos(\omega_t + \omega_{)}\sin\left[\varphi_m \Pi_{(}t)\right]\right\}.$$

Учитывая, что девиация фазы $\varphi_m = 0.5$, имеем:

$$\cos\left[\frac{\pi}{2}\Pi_C(t)\right] = \cos\left(\pm\frac{\pi}{2}\right) = 0,$$

$$\sin\left[\frac{\pi}{2}\Pi_C(t)\right] = \sin\left(\pm\frac{\pi}{2}\right) = \pm 1 = \Pi_C(t).$$

From here

$$u_C(t) = U_{mC}\Pi_C(t)\cos(\omega_C t + \varphi_C). \tag{4.49}$$

To ensure the normal operation of the traffic channel, the carrier in the synchronization channel must be restored. The simplest transformation that gives a similar effect is the squaring of the function $u_C(t)$, which can be performed, for example, with a quadratic detection. Indeed, having squared expression (4.49), we obtain [31]

$$u_C^2(t) = U_{mC}^2\Pi_C^2(t)\cos^2(\omega_C t + \varphi_C) = \frac{U_{mC}^2}{2}[1 + \cos(2\omega_C + 2\varphi_C)]. \tag{4.50}$$

Since $\Pi_C^2(t) = 1$, the modulation disappears. Thus, a filter tuned to twice the frequency $2\omega_C$ can isolate a purely harmonic waveform. After dividing the frequency of this vibration by two, the carrier is recovered.

Note that when dividing the frequency, phase uncertainty occurs in the output voltage $\pm\pi$, i.e., the sign of the recovered carrier is determined by uncontrolled initial conditions in the divider. The consequence of this is a possible "reverse operation" mode, in which the positive and negative signals at the output of the phase detector change their sign. As a result, the ones and zeros in the received information sequence are reversed.

The possibility of a random transition to the "reverse operation" mode is a characteristic disadvantage of any carrier recovery schemes when detecting modulated oscillations of the PCM-FM type. To exclude the "reverse operation" mode, special types of codes are used or the transmission of control messages is provided. In practice, to restore the suppressed carrier, circuits are used in which nonlinear signal conversion is combined with the automatic tuning of the local oscillator in a single tracking system. This eliminates the need for a special frequency divider. One of the widely used variants of such devices is the Costas circuit shown in Fig. 4.20.

The frequency of the reference local oscillator in the synchronization mode coincides with the signal frequency. The signal voltage (4.49), together with the noise, is fed to two phase detectors, the reference voltages of which are phase-shifted by $\pi/2$:

$$\left.\begin{aligned} u_{G1}(t) &= U_{m\Gamma}\sin(\omega_C t + \varphi_G), \\ u_{G2}(t) &= U_{m\Gamma}\sin\left(\omega_C t + \varphi_G + \frac{\pi}{2}\right) = U_{m\Gamma}\cos(\omega_C t + \varphi_G) \end{aligned}\right\}. \tag{4.51}$$

At the output of the phase detectors, there are sequences of positive and negative pulses corresponding to the zeros and ones of the transmitted symbols. The amplitudes of these pulses depend on the phase shift of the heterodyne and signal voltages $\Delta\varphi = \varphi_C - \varphi_G$. Moreover, due to the phase shifter in the second channel on the first phase detector, the pulse amplitude is proportional to - sin $\Delta\varphi$, and on the second - cos $\Delta\varphi$:

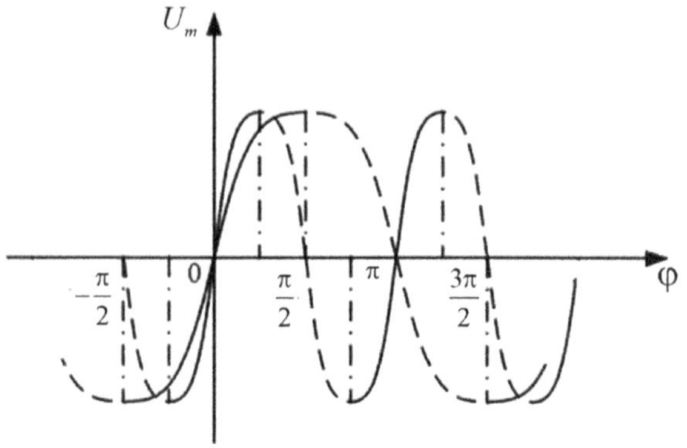

Fig. 4.21 The sign of the discriminator characteristic of the control system steepness

$$\left.\begin{aligned} u_1(t) &= K_D U_{mC} \Pi_C(t) \sin \Delta\varphi, \\ u_1(t) &= K_D U_{mC} \Pi_C(t) \cos \Delta\varphi, \end{aligned}\right\} \tag{4.52}$$

After filtering, both of these sequences are multiplied, resulting in a control voltage

$$u_P(t) = K_B K_D U_{mC}^2 \sin \Delta\varphi \cos \Delta\varphi = \frac{K_B K_D}{2} U_m^2 \sin 2\Delta\varphi. \tag{4.53}$$

In the process of multiplying with the coefficient, a nonlinear transformation of the signal occurs, removing the modulation, since $\Pi_C^2(t) = 1$.

The regulating voltage turns out to be proportional $\sin 2\Delta\varphi$, i.e., the discriminator characteristic of the tracking system has the form of a sinusoid, as in a conventional PLL system.

The BC correction block determines the required astatism order of the next contour. In its absence, the system has first-order astatism [32].

Since the discriminator characteristic of the control system is sinusoidal and proportional $\sin 2\Delta\varphi$, the sign of its steepness, which ensures stable equilibrium in the system, is periodic with an interval $k\pi$ (Fig. 4.21).

Consequently, the steady-state voltage of the reference local oscillator $u_{SG}(t)$ applied to the phase detector of the information channel can be either in-phase or antiphase with the signal, which leads to the possibility of "reverse operation".

The characteristics of the Costas circuit are very similar to the characteristics of the tracking filter based on the PLL system. There are also capture and hold bands and, therefore, frequency targeting is necessary for the system to enter synchronism at large detunings. The Costas circuit, like the PLL, has similar tracking errors due to dynamic and noise errors.

The Costas circuit is not the only possible solution for suppressed carrier recovery. There are other options that differ from it both in principle and in circuit design. In particular, Fig. 4.22 shows a variant of G.A. Travina with a limiter and a multiplier removed from the control loop, to which the signal voltage (4.49) and the limited detected voltage from the second phase detector $u_2(t)$ are applied.

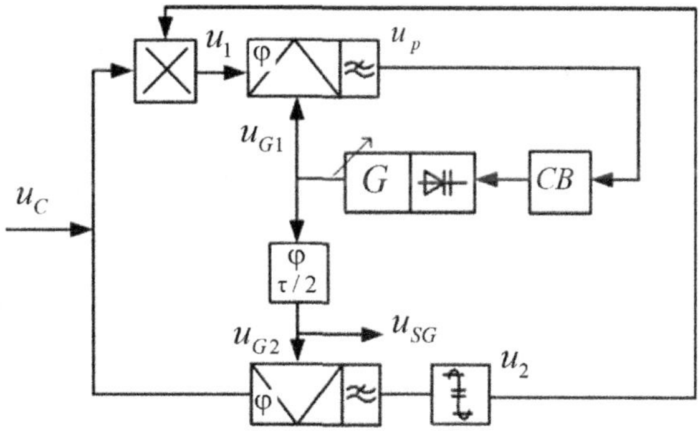

Fig. 4.22 A variant of G.A. Travina with a limiter

In accordance with expressions (4.52), it is equal to

$$u_2(t) = K_L K_D U_{mC} \Pi_C(t) \cos \Delta\varphi.$$

At the output of the multiplier, taking into account that $\Pi_C^2(t) = 1$, we have

$$u_1(t) = u_2(t)u_c(t) = K_B K_L K_D U_{mc}^2 \cos \Delta\varphi \cos(\omega_c t + \varphi_c).$$

In other words, the signal is inversely manipulated in the multiplier $u_C(t)$, which makes it possible to form an un-manipulated primary voltage, which is then filtered by the PLL system. Since the multiplication occurs at a high frequency, a conventional balanced modulator can be used to implement it.

At the output of the first phase detector, which plays the role of a phase discriminator with a reference oscillation $u_{G1}(t)$ [see formula (4.51)], we obtain the regulating voltage

$$u_{G1}(t) = K_B K_L K_D^2 U_{mC}^2 \cos \Delta\varphi \sin \Delta\varphi = \frac{K_B K_L K_D^2}{2} U_{mC}^2 \sin 2\Delta\varphi.$$

Comparing it with the voltage determined by expression (4.53), we see that it is similar to the same voltage in the Costas circuit.

With the same structure of the control circuit, both of the above carrier recovery schemes are described by the same differential equation and, of course, have the same characteristics. However, in the Travin scheme, it is structurally easier to ensure the decoupling of the multiplied voltages. It also has some possibilities for improving noise immunity by additional processing of the output signal of the first phase detector. For example, the use of an amplitude limiter in this circuit with sufficiently large signal-to-noise ratios at the input provides a decrease in the phase error of the reconstructed carrier.

4.11 Specific Features of Relative Phase Shift Keying Signals with Relative Phase Manipulation

One of the methods of eliminating the "reverse work" in π-manipulation is the use of relative phase-shift keying (RPSK), or, as it is also called, phase difference modulation, proposed by N.G. Petrovich.

With RPSK, the phase of the high-frequency filling of the transmitted elementary message depends on the phase of the previous one. So, for example, the following law of the formation of the RPSK signal can be established: the carrier phase changes abruptly by an amount at times when in the modulating sequence $\Pi_C(t)$ either the binary code symbols change from one to zero or the zero symbol is retransmitted. The timing diagram of such a signal is shown in Fig. 4.23.

Considering it, we can formulate the following algorithm for receiving signals from RPSK. If the phases of the given and the previous messages φ_{i-1} do not coincide, then a zero of the binary code (negative message) is accepted, if the phases are the same, then ones are accepted (positive message). The comparison of the parcels and the decision on the sign of the parcel can be done in various ways. The most common of these are the autocorrelation reception method and the polarity comparison method, or the coherent reception method [33].

A simplified block diagram of the autocorrelation technique is shown in Fig. 4.24. It contains a phase detector with a low-pass filter, a delay line and a threshold shaper. The phase detector compares the phases of the received signal and the signal delayed by the chip time τ_0:

$$\left.\begin{aligned} u_C(t) &= U_{mC}\cos(\omega_C t + \varphi_C + \varphi_i), \\ u_C(t) &= U_{mC}\cos\left[\omega_C(t-\tau_0) + \varphi_C + \varphi_{i-1}\right]. \end{aligned}\right\} \tag{4.54}$$

At the output of the phase detector, a voltage occurs, determined by the phase difference $\Delta\varphi = \omega_C\tau_0 + \varphi_i - \varphi_{i-1}$:

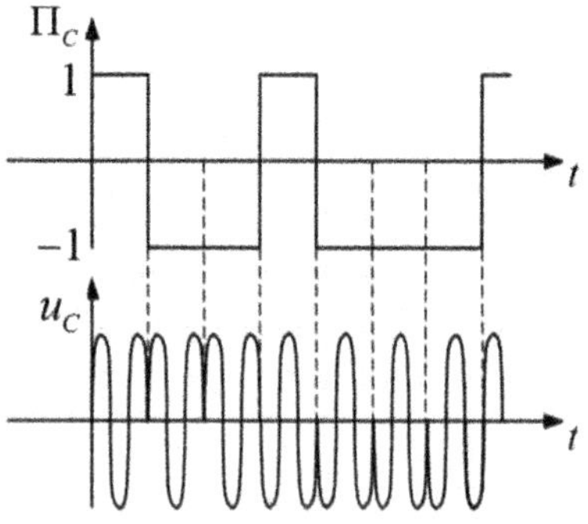

Fig. 4.23 The timing diagram of the RPSK signal

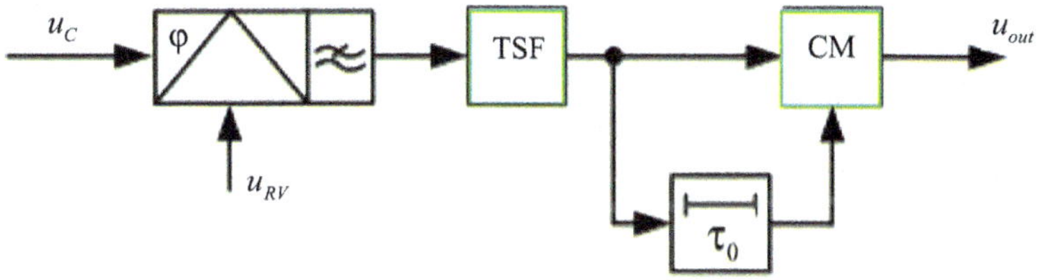

Fig. 4.24 A simplified block diagram of the autocorrelation technique

$$u_{out} = K_{PD}U_{mC}\cos\Delta\varphi = \begin{cases} K_{PD}U_{mC}\cos\omega_C\tau_0 & \varphi_i - \varphi_{i-1} = 0, \\ -K_{PD}U_{mC}\cos\omega_C\tau_0 & \varphi_i - \varphi_{i-1} = \pi, \end{cases} \quad (4.55)$$

If $\omega_1\tau_0 = 2k\pi$, then the output voltage is maximum and equal to $\pm k_{PD}U_{mC}$. Since in autocorrelation reception a delayed signal is used as a reference voltage, the appearance of "reverse operation" is fundamentally excluded.

With the coherent method of reception, the signal is detected by a conventional phase detector, during normal operation of which a reference voltage is formed by one of the methods discussed in Sect. 4.10.

The decision on the nature of the received premise, as can be seen from the structural diagram shown in Fig. 4.24, is received by the polarity comparison device (multiplier of the signs of the joint venture), to which, from the output of the threshold-forming (TSF) device i, the following and delayed τ_0 for a while τ_0 are sent:

$$u_{out\,i}(t) = u_{B\,i}(t)u_{B\,i-1}(t).$$

Obviously $u_{out\,i}(t)$, it is positive if the messages have the same polarity, which corresponds to the reception of one of the binary codes. With different polarities of the parcels $u_{out\,i}(t)$, it is negative and a negative pulse appears at the output, corresponding to zero.

With a coherent method of reception, a phase jump π in the synchronization device will cause a change in the sign of the i-th transmission $u_{Bi}(t)$, which will lead to the appearance of "reverse operation". However, after a time interval τ_0, the message $u_{Bi-1}(t - \tau_0)$ also changes its sign and the correct registration of the transmitted message is restored. Thus, a jump in the phase of the reference oscillation results in only a single error. In a properly designed system, jumps in the phase of the reference voltage occur relatively rarely and practically have little effect on the noise immunity of the reception.

References

1. Belousov AP(1959) Calculation of the noise figure of radio receivers. –M.: State. publishing house of defense. Prom-sti, 135 p
2. Bogdanovich BM (1980) Nonlinear distortions in receiving and amplifying devices. –Moscow: Svyaz, 280 p
3. Buga NN, Fal'ko AI, Chistyakov NI (1986) Radio receiving devices: textbook. for universities (NI Chistyakova ed). –M.: Radio and Communication, 428 p
4. Van der Ziel (1973) Noise (sources, description, measurements): translated from English. –M .: Sov. Radio, 178 p
5. Voishvillo GV(1983) Amplifying devices. –M .: Radio and Communication, 264 p
6. Vorobiev VI (1983) Optical location for radio engineers. –M.: Radio and Communication, 176 p
7. Goldenberg LM, Matyushkin BD, Polyak MN (1985) Digital signal processing: a handbook. –M.: Radio and Communication, 312 p
8. GOST 27002-83. Reliability in technology. Terms and definitions
9. GOST 93611-79. Electromagnetic compatibility of radio electronic means. Terms and definitions
10. Gutkin LS, Chentsova OS (1958) Transient processes in the system: high-frequency amplifier—detector. Radio Eng (11):18–23
11. Ditkin VA, Prudnikov AP (1965) A guide to operational calculus. –M.: Higher. shk., 466 p
12. Evtyanov SI (1948) Transient processes in receiving-amplifying circuits. –M.: Radio and Communication, 210 p
13. Zernov NV, Karpov VG (1972) Theory of radio engineering circuits. Leningrad: Energiya, 816 p
14. Vasilchenko NV, Borisov VA, Kremenchug LS, Levin GE, Kurbatova LN, Vasilchenko NV (ed) (1983) Measurement of parameters of optical radiation receivers. –M.: Radio and communication, 320 p
15. Kanevsky ZL, Finkelstein MI (1963) Fluctuation interference and detection of pulsed radio signals. –M.; L.: Gosenergoizdat, 243 p
16. McElroy JH (1977) Communication systems for near space using CO_2 lasers. TIIER (2):54–89
17. Malashin MS, Kaminsky RP, Borisov YB (1983) Basics of designing laser locating systems: textbook manual for universities – M.: Higher. shk., 207 p
18. Martynov VA, Selikhov YI (1980) Pan-frame receivers and spectrum analyzers (GD Zavarin Ed). –M.: Sov. Radio, 352 p
19. Pervachev SV (1982) Radioautomatics: textbook for universities. –M .: Radio and communication, 296 p
20. Pestryakov VB, Kuzenkov VD (1985) Radio engineering systems. –M.: Radio and communication, 376 p
21. Poberezhsky ES (1987) Digital radio receivers.- M.: Radio and communication, 184 p
22. P r a t V.K (1972) Laser communication systems: translated from English. –M.: Svyaz, 232 p
23. Golubkov AP, etc ., MA Sokolov (Ed) (1984) Design of radar receiving devices: textbook for radio engineering special universities. – M.: Higher school, 335 p
24. Bankov VN, Barulin LG, Zhodzishsky MI and others; Barulina LG (ed) (1984) Radio receivers. – M.: Radio and communication, 272 p
25. Zyuko AG, Korobov YF, Levitan GI, Simontov IM, Falco AI; AG Zyuko (ed) Radio receivers (1975) – M.: Communication, 400 p
26. Teplyakov IM, Roshchin BV, Fomin AI, Weizel VA; Teplyakova IM (ed) (1982) Radio systems for transmitting information: textbook for universities. – M.: Radio and communication, 264 p
27. Ross M (1969) Laser receivers: translated from English (AV Nevsky ed), Moscow: Mir, 520 p
28. Korostelev AA, Klyuev NF, Miller YA and others, Dulevich VE (ed) (1978) Theoretical foundations of radar: textbook for universities. – M.: Soviet Radio, 608 p
29. Xyuz RS(1980) Logarithmic amplifiers. Moscow: Sov. Radio, 352 p.

30. Zhodzishsky MI, Sila-Novitsky SY, Prasolov VA and others, Zhodzishsky MI (ed) (1980) Digital phase synchronization systems. – M.: Soviet Radio, 208 p
31. Chistyakov NI (1938) To the calculation of one discriminator circuit. News of the electrical industry of a weak current. No. 2
32. Sheremetyev AG (1971) Statistical theory of laser communication. – Moscow: Communication, 264 p
33. Vimberg GP, Vinogradov YV, Fomin AF and others, Zenkiewicz OA (ed) (1972) Energy characteristics of space radio lines. – M.: Sov. Radio 436 p

Chapter 5
Range Radio Receiving Devices

5.1 General Information About Range RAN

Range radio receivers (RRR) are designed to receive radio signals in the frequency range exceeding their bandwidth. The RRR includes, for example, the overwhelming majority of receivers of communication systems, receivers of radio engineering systems using multi-frequency signals, panoramic, measuring radio control systems, etc. The main tasks that can be solved by RRR are [1]

1. Reception of signals with known and unknown parameters.
2. Monitoring the operation of a large number of individual radiation sources.
3. Separation of signals from any of the radio emission sources from the set of signals entering the RRR input.
4. Measurement of parameters of received signals.
5. Assessment of the electromagnetic environment at the location of the radio receiver, etc.

The structural diagrams of band radio receivers are largely determined by the method of analyzing the frequency range and the purpose of the radio receiver. Distinguish between methods of simultaneous (parallel), sequential and combined analysis or viewing the frequency range.

In a parallel analysis, all electromagnetic radiation of various origins, located in a certain frequency band, are detected and analyzed simultaneously, without missing them, which is a very significant advantage of this method of constructing a RRR. A possible simplified block diagram of such a receiver is shown in Fig. 5.1. Here, the preselector "I" forms a broadband RRR path and has a bandwidth that covers the entire frequency range in which the receiver operates. In the linear path, depending on the purpose of the RRR, there can be several frequency conversions. Band pass filters 3 cover the entire analyzed range and serve for frequency separation of signals. From the outputs of the amplifiers 4, the signal enters the analyzing devices 5, the structure of which depends on the purpose of the RRR.

S. V. Dvornikov et al., *Radio-Electronic Equipment in Civil Aviation*, Springer Aerospace Technology, https://doi.org/10.1007/978-981-19-6199-1_5

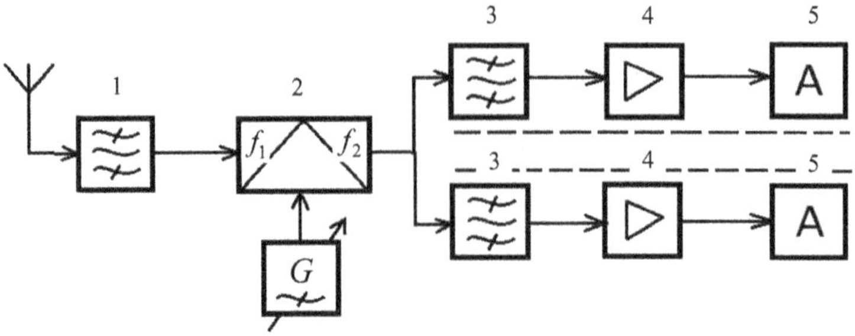

Fig. 5.1 A possible simplified block diagram of an RRR receiver

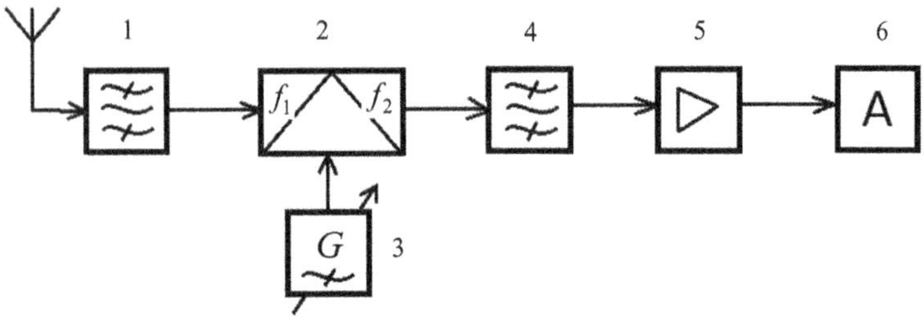

Fig. 5.2 A simplified block diagram of a receiver, when the tuning is carried out in the frequency range and the signals are detected sequentially, with a broadband preselector

In radio receiving devices for sequential analysis, the tuning is carried out in the frequency range and the signals are detected sequentially. Note that, in this case, there may be a skipping of signals, especially short-term ones. A simplified block diagram of a receiver of this type with a broadband preselector is shown in Fig. 5.2.

Here, the preselector "I" has a bandwidth equal to the analyzed frequency band, and the local oscillator 3 provides the RRR tuning in a given frequency band. The law of its restructuring can be different. After the selection of the signal in the narrowband path by the filter 4 and its amplification by the amplifier 5, the signal enters the analyzing device 6.

Combined frequency analysis is based on the use of parallel and sequential analysis methods in RRR, which makes it possible to improve their basic characteristics.

The method of sequential search, detection and reception of signals is implemented in receiving devices for various purposes, for example, in radio control systems of radio engineering systems and communication systems, measuring and panoramic radio transmitters, spectrum analyzers, etc., and the method of simultaneous analysis is most often used in measuring and panoramic radio transmitters [2].

5.2 Main Qualitative Indicators of Range RPU

A number of qualitative indicators are introduced to compare RRRs for various purposes and assess the degree of compliance with their requirements. The main ones, along with sensitivity, selectivity, dynamic range, noise immunity, etc., include some specific indicators: swath, speed of view, resolution, etc. The latter mainly characterize the operation of measuring and panoramic radio control systems.

The span Φ_0 determines the frequency range within which the RRR receives and analyzes signals:

$$\Phi_0 = f_{\text{c max}} - f_{\text{c min}}, \tag{5.1}$$

where $f_{\text{c max}}$ and $f_{\text{c min}}$, respectively, are the maximum and minimum frequencies of the analyzed frequency range.

In parallel analysis receivers, the span is determined by the bandwidth of the broadband path (preselector) or the total bandwidth of the receiver's narrowband channels (if it is less than the bandwidth of the broadband path).

In sequential analysis receivers, the value Φ_0 is determined by the preselector bandwidth and the possible range of variation of the local oscillator frequency and the tuning of the preselector if it is tuned [3].

The sweep speed measures how quickly the swath is scanned Φ_0. Parallel analysis provides simultaneous, instant viewing of the strip Φ_0. In sequential analysis, the scan rate is determined by the rate of change of the local oscillator frequency:

$$\gamma_0 = \frac{\Phi_0}{T_c}, \tag{5.2}$$

where T_c is the time of changing the local oscillator frequency from the minimum to the maximum value (or vice versa), at which the band is scanned Φ_0. Sometimes this value is called the scan period, which is related to the scan rate F_c by the ratio $T_c = 1/F_c$.

Resolution determines the ability of the receiver to resolve (separate) two signals adjacent in frequency. Quantitatively, it is estimated by the minimum frequency interval between adjacent signals, at which they can be received and analyzed separately.

The resolution in parallel analysis is determined by the shape of the amplitude-frequency characteristics of the channels of the narrowband path and their bandwidth. This is illustrated in Fig. 5.3, which shows the normalized frequency response of the four channels.

If the amplitude-frequency characteristics of adjacent narrowband channels are adjacent to each other at some level, for example, at a level of 0.7, then due to the overlap of the amplitude-frequency characteristics, the signal at the frequency f_1 will cause a response at both the output of the first and the output of the second channels, which can lead to false triggering of the second channel. It is obvious that the level

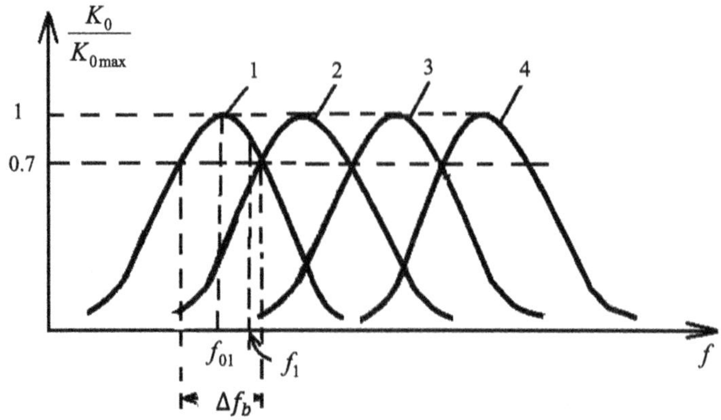

Fig. 5.3 The normalized frequency response of the four channels

of the false signal in the second channel will be the less, the greater the squareness of the amplitude-frequency characteristics of the channels [4, 5].

The accuracy of measuring the frequency of received signals in parallel analysis, if there is no special frequency meter, is determined by the bandwidth Δf_b of the narrowband channel. Indeed, any signal with a frequency, for example, f_1 falling into this (first) channel, is assigned the average frequency of the bandwidth of this channel f_{01}. Then there will be an error in the measurement of the frequency $\Delta f = \frac{1}{2}\Delta f_b$. False triggering of the second channel can be excluded by using a threshold device in it.

The resolution in sequential analysis is determined by the dynamic amplitude-frequency characteristic of the receiver, which is understood as the dependence of the voltage at its output on the frequency at a given rate of change of the receiver tuning frequency.

The phenomenon of the dynamic effect is due to the fact that when the receiver is tuned, the time spent by the received signal in the passband of the receiver may turn out to be less than the settling time of the signal caused by transient processes. This leads to the fact that the maximum of the signal at the output of the receiver does not appear at the moment of tuning the receiver exactly to the frequency of the received signal, but somewhat later, which will correspond, as it were, to the frequency offset of the amplitude-frequency characteristic. This shift is the greater, the greater the tuning rate of the radio receiver [2]. In addition, the maximum of the dynamic amplitude-frequency characteristic decreases compared to the maximum of the static characteristic, the bandwidth of the receiver (dynamic bandwidth) Δf_{Db} increases, which leads to a deterioration in the resolution and sensitivity of the receiver, the amplitude-frequency response becomes asymmetric, and the slope of the voltage drop exceeds the slope of the rise. The type of dynamic characteristics depending on the tuning rate of the receiver is shown in Fig. 5.4.

Quantitatively, the apparent bandwidth expansion is characterized by the expression

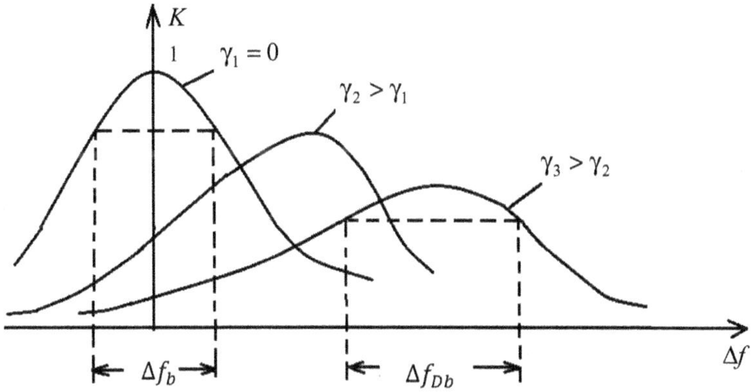

Fig. 5.4 The type of dynamic characteristics, depending on the tuning rate of the receiver

$$\frac{\Delta f_{Db}}{\Delta f_b} = \sqrt{1 + \left(\frac{K\gamma}{\Delta f_b^2}\right)^2}. \tag{5.3}$$

Here K is a coefficient that depends on the shape of the amplitude-frequency characteristic and γ is the speed of tuning the radio receiver in the passband.

From relation (5.3), it follows that the apparent bandwidth expansion can be neglected if $\Delta t \Delta f_b \neq 1$, where $\Delta t = \frac{\Delta f_b}{\gamma}$ is the receiver tuning time by an amount equal to its static bandwidth.

The dynamic bandwidth depends on the shape and bandwidth of the static frequency response of the narrowband channel and the sweep rate of the receiver. Therefore, in sequential analysis, to obtain the best resolution, it is necessary to provide a certain relationship between the static system bandwidth and the rate of change of the receiver tuning frequency. There is an optimum static bandwidth $\Delta f_{b\text{opt}}$ at a given sweep rate γ where the minimum dynamic bandwidth can be obtained. This conclusion can be explained as follows. The dynamic bandwidth of the electoral system cannot be less than the static one and depends on both the static bandwidth Δf_b and the tuning rate γ. At a given value γ, the decrease in Δf_b leads to a decrease in Δf_{Db} only up to a certain limit, after which the expansion Δf_{Db} due to transient processes in electoral systems becomes predominant. The latter mainly determines the meaning Δf_{Db}. Further narrowing leads to an increase in Δf_{Db}.

It should be noted that both in parallel and in sequential analysis, the resolution of the DRPU depends not only on its selectivity characteristics, but also on the ratio of the amplitudes of the analyzed signals. This is due to the fact that a strong signal, for example, in parallel analysis, can cause a response at the output of two adjacent receiver channels at the same time, exceeding the threshold value corresponding to the output voltage at a normal input signal level and falling into the channel bandwidth. This will cause false triggering of the channels and exclude the possibility of signal resolution.

5.3 Methods of Adjustment of DRPU and Division of Its Operating Frequency Range into Sub-Bands

In range radio receivers of sequential analysis, they must be tuned to any of the frequencies that are within the operating range, i.e., in the range from $f_{c\ min}$ to $f_{c\ max}$. This is achieved in the most general case due to the use in the circuits of a preselector, a local oscillator and sometimes an IF amplifier, which makes it possible to change their natural frequency. Such elements can be variable capacitors, inductors, varicaps, etc. In this case, the restructuring of the receiver can be

- smooth, carried out by smoothly changing the natural frequency of tuning the tunable electoral systems;
- discrete, occurring due to abrupt switching of the elements of all tunable electoral systems or smooth changes in the elements of the electoral systems of the preselector and the use of a grid of discrete frequencies of the local oscillator, as a result of which reception is provided in signals at discrete fixed frequencies;
- combined, occurring in jumps during the transition from one section of the operating frequency range to another and smoothly—within each section.

A smooth adjustment can be carried out by either mechanical or electronic change in the value of the capacitance or inductance of the circuits. The choice of this or that method is carried out on the basis of the requirements for the rate of restructuring or from those advantages and disadvantages that are inherent in the indicated restructuring methods [6].

The restructuring time of the radio receivers (RR) when using mechanical methods is tens of seconds, and electronic is fractions of seconds.

Mechanical adjustment is usually carried out using a variable capacitor. This achieves a comparative ease of obtaining the required law of frequency change and the possibility of realizing a large overlap factor of the frequency range.

In electronic tuning, varicaps are most commonly used, the use of which makes it possible to obtain a high speed and ease of tuning by changing the control voltage or current, a relatively high stability of the tuning frequency when external conditions change, and the insensitivity of the tuning system to vibrations.

However, varicaps also have significant drawbacks, such as capacitance nonlinearity (which, at high levels of interfering signals at the receiver input, can lead to a deterioration in multi-signal selectivity), the need for high installation accuracy and stability of the control voltage and a small frequency band overlap coefficient.

Discrete tuned receivers use stores of capacitors C or L inductors (or both at the same time). Switching control L and C is carried out with the help of special electronic key circuits on reed switches. The restructuring time of circuits with discrete L and C with the help of reed switches is tenths and the date is hundredths of a second.

In the case when the swath, i.e., the entire range of received frequencies from $f_{c\ min}$ to $f_{c\ max}$ cannot be covered with tunable elements, or when high requirements are imposed on the constancy of its main indicators and frequency accuracy of tuning within the operating frequency range, the frequency range of the radio receiver is divided into subbands.

So, if the receiver range coefficient is greater than the possibilities for tuning electoral systems, for example, with capacitance or inductance, i.e., corresponding subband coefficients

$$K_{D\ f} = \frac{f_{c\ max}}{f_{c\ min}}, \tag{5.4}$$

$$K_{DC} = \sqrt{\frac{C_{max}}{C_{min}}} < K_{D\ f} \text{ or } L_{DC} = \sqrt{\frac{L_{max}}{L_{min}}} < K_{D\ f},$$

where C_{max} and C_{min}, L_{max} and L_{min} are the maximum and minimum possible values of capacitance and inductance that can be obtained in the circuit using the variables L and C, then the entire frequency range of the RR is divided into subranges. In this case, the maximum value of the subband coefficient when tuning by a variable capacitor is usually in the range of 2.5–3.5; variable inductance coil—within 1.4–2.0 and varicap—in the range of 1.3–1.5.

There are three main ways of dividing the total frequency range into subbands: with a constant subband K_D gain for all subbands, with a constant sweep interval for each subband and using a combination of the first two methods.

The first method of dividing into subbands, in which $K_{D1} = K_{D2} = \ldots = K_{DN}$, has the significant disadvantage (for example, in connected radio transmission systems) that in different subbands the frequency density of stations per unit length of the scale turns out to be different (it turns out to be the largest in the subbands corresponding to the upper tuning frequencies). This leads to an inaccurate setting of the tuning frequency of the RRR. In receivers with discrete tuning, the frequency is set using electronic-digital dial devices, and, in this case, the accuracy of setting the receiver tuning frequency is determined only by the accuracy of the local oscillator frequency. The advantages of this method are the simplicity of the implementation of the preselector and local oscillator circuits and the smallest number of subbands.

When dividing the frequency range into subbands according to the second method, the same frequency density is ensured on all subbands, which gives the same frequency setting accuracy in each subband; however, the number of subbands turns out to be greater than in the first method, and, in addition, the preselector and local oscillator circuits become more complex.

The third method uses the merits of both the first and second methods. Namely, the lower part of the range is divided into subbands with equal K_D, and the upper with a constant interval of tuning frequencies for each subband.

5.4 Features of Structural Schemes of DRPU

The principles of constructing structural diagrams of the RRR depend on the purpose of the RRR and the method of viewing and analyzing the frequency range implemented in them. For modern range radio transmitters, especially with sequential viewing of the frequency range (measuring and communication radio transmitters), the use of multiple frequency conversion is characteristic, which makes it possible to implement high, often contradictory, requirements for them [7].

The broadband path in the RRR determines such indicators as single and multi-signal selectivity, dynamic range and receiver sensitivity. In the broadband path, in addition to the blocking effect, cross-distortion and intermodulation between the useful signal and powerful interference, it is necessary to take into account the interaction between the components of the group signal entering the input of the broadband path and leading to the appearance of the so-called nonlinear noise (intermodulation of the second and third order). The intensity of this noise depends on the parameters of the nonlinearity of the electronic devices, the bandwidth and the amplification of the stages of the broadband path. An increase in the level of nonlinear noise leads to a deterioration in the resulting signal-to-noise ratio at the output of the radio receiver, i.e., to a decrease in the sensitivity of the receiver, a deterioration in its resolution and a decrease in such an important indicator of radio receivers as the probability of correct detection of signals.

Since the level of interference during multi-signal exposure is largely determined by the structure and parameters of the broadband path, there is such a combination of the structure and distribution of the path parameters, in which a real broadband path provides minimal degradation of the signal-to-noise ratio in comparison with an ideal linear path. The optimal distribution of gain between the stages of a wideband path is important for improving the signal-to-noise ratio, since the gain of the stages has a contradictory effect on the signal-to-noise ratio. So, for example, with an increase in the gain, the signal-to-noise ratio due to the intrinsic (linear) noise of the stages decreases, and the intensity of nonlinear noise increases.

In addition to the optimal gain distribution, the signal-to-noise ratio is largely influenced by the selectivity distribution in the broadband path, the number of stages and the nonlinearity parameters of electronic devices (see Sect. 1.4).

The circuit design of the elements of the broadband path of the RRR does not differ from the known elements of the corresponding frequency range, built taking into account their broadband [8]. It should be borne in mind that often the receiver span is significant and it becomes impossible to implement elements of a broadband path with the required characteristics, for example, bandwidth, selectivity and gain. Then the block diagrams of the receiver with parallel and sequential analysis change and take the form shown in Figs. 5.5 and 5.6. In parallel and sequential analysis, here the broadband path is covered by a set of filters and amplifiers, the middle frequencies of the passbands of which for adjacent filters differ from each other by the amount of their passband.

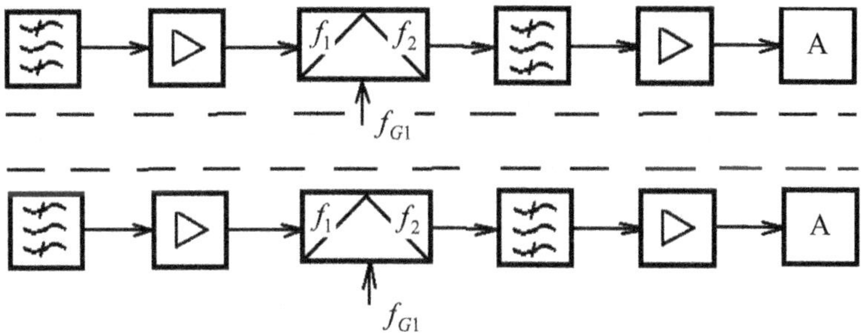

Fig. 5.5 The block diagrams of the receiver with parallel analysis

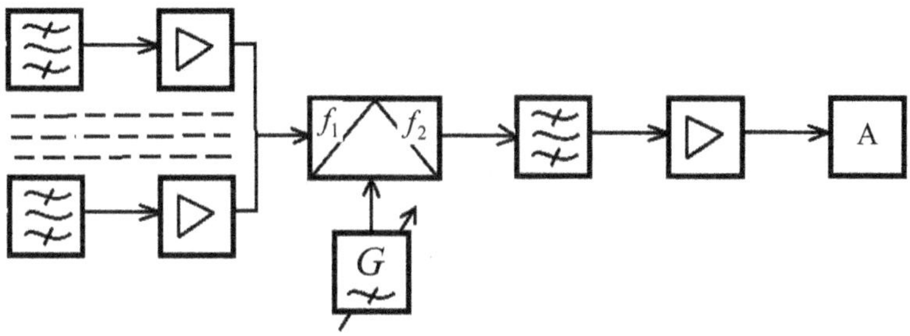

Fig. 5.6 The block diagrams of the receiver with sequential analysis

In parallel analysis, one filter of a broadband path can be connected simultaneously to several narrowband paths if the bandwidth of a broadband amplifier overlaps several bands of narrowband filters. In receivers of sequential analysis (Fig. 5.6), if the local oscillator cannot provide smooth tuning over the entire range of the broadband path, i.e., in the span, then combined tuning is used: smooth—when analyzing signals in the range of each of the filters of the broadband path and discrete—when switching from filter to filter.

The block diagram of the narrowband RRR path depends not only on the method of viewing the frequency range, but also on the requirements for the receiver.

In receivers of simultaneous analysis, in which the heterodyne frequencies are fixed, and, therefore, the intermediate frequencies are also constant, the structure of the narrowband path in terms of the number of conversions and circuit implementations of the IF amplifier paths is determined mainly by the method of resolving the contradiction between high selectivity in an adjacent channel and selectivity in side reception channels (see Sect. 1.2).

In receivers of sequential analysis (reception), two or more frequency conversions, in addition to the indicated reasons, can be caused by the choice of the structural diagram of the receiver from the condition of fulfilling the requirements for frequency accuracy. In particular, two frequency conversions are carried out in a DRPU with a single-band first local oscillator, with a quartz-crystal first and a smooth second local oscillator, as well as in an RRR, in which the principle of compensation of the frequency drift of an auxiliary local oscillator is used, which is a common source in the formation of heterodyne frequencies, etc.

It should be noted that various frequency conversion options are used in the range RRRs: conversion with spectrum transfer of the received signal both below the minimum and above the maximum frequency of the receiver range; sum or difference transformation and differential conversion at high or low local oscillator setting. Each of these transformations has advantages and disadvantages [9, 10].

The transfer of the signal spectrum below the minimum frequency simplifies the RFA scheme, since the number of frequency transformations is reduced, the implementation of high selectivity along the adjacent channel is simplified, but it becomes more difficult to obtain a high degree of attenuation of the side reception channels at the mirror and intermediate frequencies. Conversion of the signal spectrum above the maximum frequency of the received frequency range, which most often occurs in the longer wavelength part of the radio range, makes it possible to obtain a significant increase in the degree of suppression of side reception channels, but this conversion method increases the number of conversions.

The total transformation used when transferring the signal spectrum above the maximum frequency of the receiver range allows you to set the frequency of the first local oscillator relatively low, but this dramatically increases the number of combination side channels in the operating frequency range [7, 10].

With the difference conversion, it is necessary to take into account that at the lower setting of the local oscillator, the spectrum of the received signal after conversion is not inverted, and at the upper setting it is inverted. This must be taken into account especially when receiving signals with an unbalanced spectrum, such as single sideband signals. In the case of the upper setting of the RRR local oscillator, it is advisable to carry out the transfer of the signal spectrum "up", in the case of the lower—"down".

Let us consider, for example, some of the possible options for constructing structural diagrams of range RRs with sequential analysis. Most often, such circuits are implemented in communication and measuring receivers.

The block diagram of the radio frequency amplifier (RFA) with a smooth single-band first local oscillator is shown in Fig. 5.7. The main feature of the circuit is the use of a single-band unswitched first local oscillator 1. Local oscillator 1 has a smooth tuning in a constant and fixed range. To form the first intermediate frequency, the harmonics of this local oscillator are used, separated by a frequency multiplier. To reduce the number of multiplication steps and the nominal values of the first intermediate frequency, the upper and lower settings of the local oscillator 1 are used in relation to the operating frequencies of the subband.

The use of a single-band local oscillator in combination with parametric frequency stabilization allows in this circuit to obtain the value of the relative instability of the local oscillator frequency up to 10^{-5}. When implementing this scheme, the boundaries of the subbands are rigidly determined by the range of the reference generator. With an increase in the number of the used harmonic of the local oscillator frequency, the frequency interval of the subband increases $\Delta f_{SI} = N(f_{G1\ \mathrm{max}} - f_{G1\ \mathrm{min}})$. Therefore, all subbands are found to have different overlap coefficients of the frequency

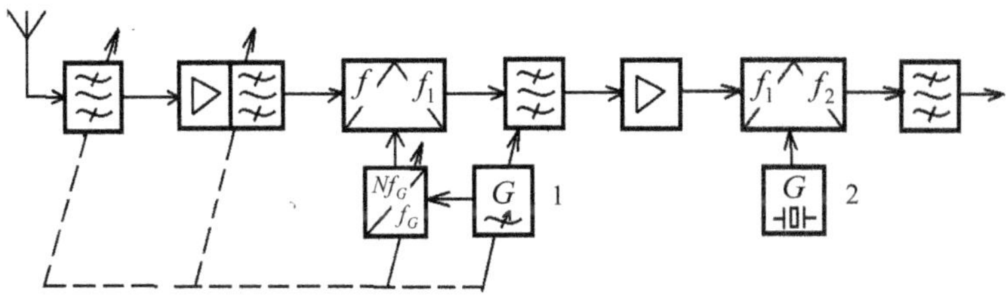

Fig. 5.7 The block diagram of the RFA with a smooth single-band first local oscillator

range and differ in the unequal frequency intervals. In the considered RFA circuit, each subband (or several of them) has its own nominal value of the first intermediate frequency and its own frequency of the second local oscillator, which leads to the need to switch over the ranges of the selective elements not only of the preselector, but also of the path of the first intermediate frequency, as well as the frequency of the second local oscillator.

It should be borne in mind that in the considered RR, when the preselector is rebuilt, its selectivity changes (and this affects the degree of suppression of side reception channels) and the nominal value of the first intermediate frequency [11].

The block diagram of a radio control system with a smooth single-band first local oscillator and a bias generator is shown in Fig. 5.8. In this scheme, the tuning range of a special smooth generator 1 is selected equal to the frequency interval in the subbands into which the range of received frequencies is divided. In a special mixer 2, with the help of discrete frequencies coming from a quartzized bias oscillator, the frequency range of oscillator 1 is transferred to the required range of variation of the local oscillator frequency f_{G1}, which is allocated by amplifier 4. In the circuit under consideration, the first intermediate frequency is constant, and therefore there is no need to switch the local oscillator frequencies 5. Depending on the selected subband in the RFA, implemented according to this scheme, it is necessary to switch the selective elements of the preselector and the frequencies of the dummy generator.

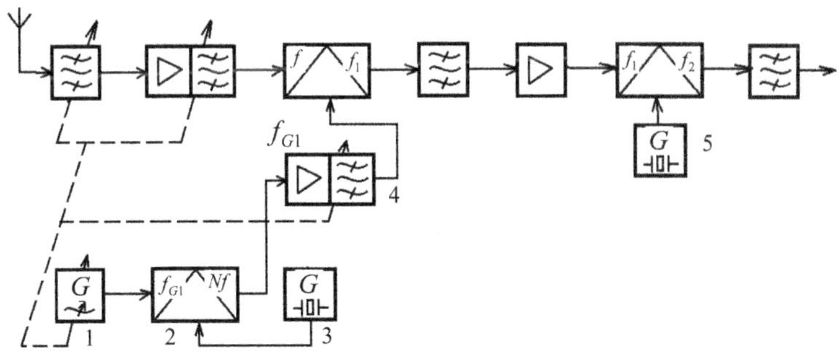

Fig. 5.8 The block diagram of a radio control system with a smooth single-band first local oscillator and a bias generator

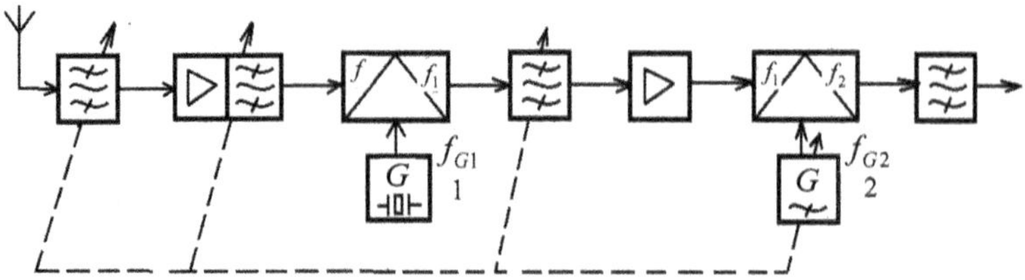

Fig. 5.9 The block diagram of an RR with a quartz-crystal first local oscillator and a smooth second one

The block diagram of an RR with a quartz-crystal first local oscillator and a smooth second one is shown in Fig. 5.9. A feature of this circuit is the constancy of the frequency of the first quartz-locked local oscillator 1 within each subband. In this case, two variants of the circuit of the first local oscillator are possible [12]:

- a multi-crystal stabilization circuit, when each subband has its own quartz, and the local oscillator circuit is switched, which worsens the stability of the local oscillator;
- a single-crystal stabilization circuit, when there is one reference crystal oscillator, and the frequencies of the first local oscillator required for each subband are obtained by frequency multiplication and division of the reference crystal oscillator (the frequency multiplication and division circuit is not shown in Fig. 5.9).

In such a coin scheme, the frequency stability of the first local oscillator is of order 10^{-5}–10^{-6}.

Due to the constancy in each subband of the frequency of the first local oscillator, the first intermediate frequency is variable and smoothly changes within the same limits when tuning in all subbands. In such receivers, the first intermediate frequency is selected high enough to provide attenuation on the first image channel. They carry out a simultaneous restructuring of the preselector, the first intermediate frequency path and the single-band second local oscillator. However, there are options for constructing a block diagram of the radio control system, when in the circuit shown in Fig. 5.9, the amplifier of the first intermediate frequency does not rebuild when the preselector tuning frequency is changed. In this case, the bandwidth of the IF amplifier should be such that it can pass all possible values of the first intermediate frequency, i.e., must equal the width of the sub-range. To ensure good selectivity in the second image channel, it is necessary to take a high second intermediate frequency, but, in this case, for the required attenuation of the adjacent receive channel, it becomes necessary to use a third frequency conversion.

The block diagram of the radio control room with a reference crystal oscillator and a smooth first local oscillator is shown in Fig. 5.10. The receiver implements the principle of compensation for the frequency deviation of the first local oscillator 1. Indeed, the deviation from the nominal frequency of the first local oscillator affects equally both the converted frequency of the received signal (after the first conversion)

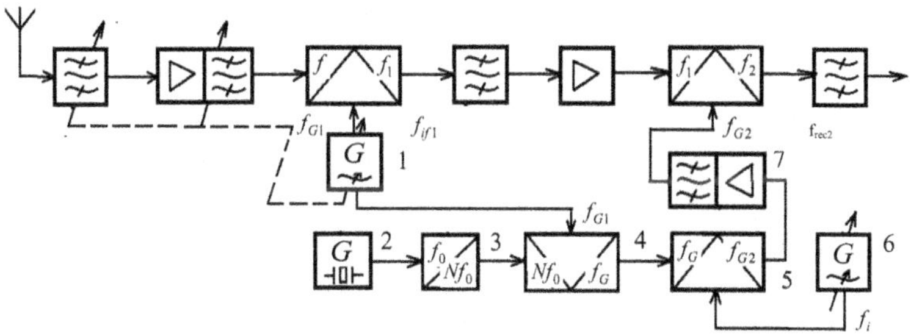

Fig. 5.10 The block diagram of the radio control room with a reference crystal oscillator and a smooth first local oscillator

and the converted harmonic frequency of the reference crystal oscillator 2 (after converting the harmonic frequency in mixer 4). During the second conversion of the signal frequency in the mixer 7, the difference of these converted frequencies is obtained and the frequency instability of the first local oscillator is compensated for.

For example, let the frequency of the first local oscillator change to Δf_G and become equal to $f_{G1} = f_{G1\,n} + \Delta f_G$. Then the first intermediate frequency will also change to Δf_G. At the top setting of heterodyne (G1)

$$f_{if1} = f_{G1} + f_C = f_{G1\,n} + \Delta f_G - f_C.$$

The frequency of the second local oscillator is formed by converting the harmonics of the reference crystal oscillator and the frequency of the first local oscillator in the mixer 4 and subsequent conversion in the mixer 5 together with the frequency of the interpolation oscillator 6. The interpolation oscillator allows you to change the frequency of the oscillations stabilized by the reference crystal oscillator without any effect on the frequency of the reference oscillator. The limits of the frequency variation of the interpolation generator correspond to the frequency interval per subband.

Thus,

$$f_{G2} = f_{G1} - Nf_0 - f_i = f_{G1\,n} + \Delta f_G - Nf_0 - f_i,$$

and the second intermediate frequency

$$f_{if2} = f_{G2} - f_{if1} = f_{G1\,n} + \Delta f_G - Nf_0 - f_{G1\,n} - \Delta f + \Delta f_C = f_C - Nf - f_i.$$

In order for the interpolation generator to have little effect on the frequency stability f_{if}, it is necessary to satisfy the inequality.

The block diagram of the radio control room with a block of reference frequencies is shown in Fig. 5.11. A feature of the circuit is the use of a grid of operating frequencies and compensation of the frequency drift of the first local oscillator. The formation of a grid of operating frequencies is as follows. The frequency of the

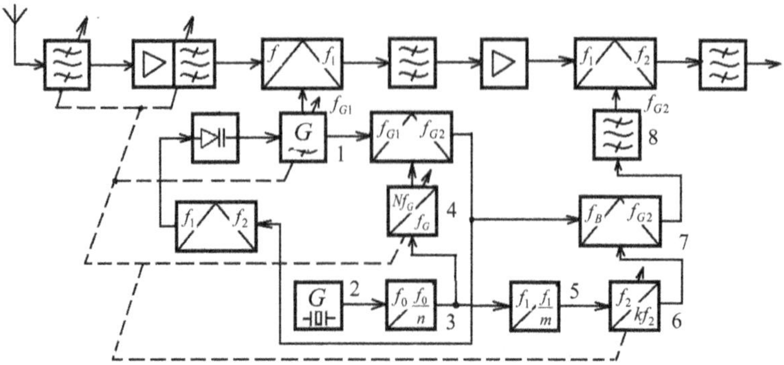

Fig. 5.11 The block diagram of the radio control room with a block of reference frequencies

reference crystal oscillator 2 is divided by the divider 3 in the required number of times in order to obtain frequency components for synchronizing the pulse generators, the signal harmonics of which are used to multiply the frequency.

In Fig. 5.11, pulse generators are combined with frequency multipliers 5 and 6. As a result of frequency division in dividers 3 and 5, and subsequent frequency multiplication in multipliers 4 and 6, "coarse" and "fine" reference frequency f_G grids are formed. The frequencies of the "fine" grid are located at predetermined intervals equal to the spacing between the fixed tuning frequencies of the RRR. The coarse grid frequency set is required to convert the frequency of the infinitely variable first local oscillator to a lower auxiliary frequency. The nominal value of the latter is chosen so that in the mixer 7 of the reference frequency unit, the conversion of its frequency and the frequencies of the "fine" grid would allow obtaining a voltage f_{G2} with a frequency that is necessary for the second stage of conversion. Filter 8 selects the required frequency f_{G2}.

As can be seen from the diagram, the principle of compensation of the frequency drift of the first local oscillator is used here, similar to that considered in the previous diagram.

Despite the possibility of compensating the frequency drift of the first local oscillator, the absolute drift of its frequency should not exceed the value at which the frequency f_{G2} will go beyond the bandwidth of filter 8. To reduce the absolute value of the frequency drift in the reference frequency block, an automatic local oscillator frequency control system is used. In a radio receiver with a block of reference frequencies, the accuracy of the frequency setting is determined by the accuracy of the nominal value of the reference crystal oscillator. Under these conditions, the simplest subband division with equal subband coefficients is used [6, 8].

A further improvement of the RRR reference frequency block is the use of a 10-day frequency synthesizer in its composition—a device that allows you to obtain any discrete local oscillator frequency by digital dialing. In this case, the structural diagrams of the broadband and narrowband paths remain the same as in Fig. 5.11.

The block diagram of a decade-long synthesizer is quite complex. However, the principle of building 10-day synthesizers can be simplified as follows (Fig. 5.12). A

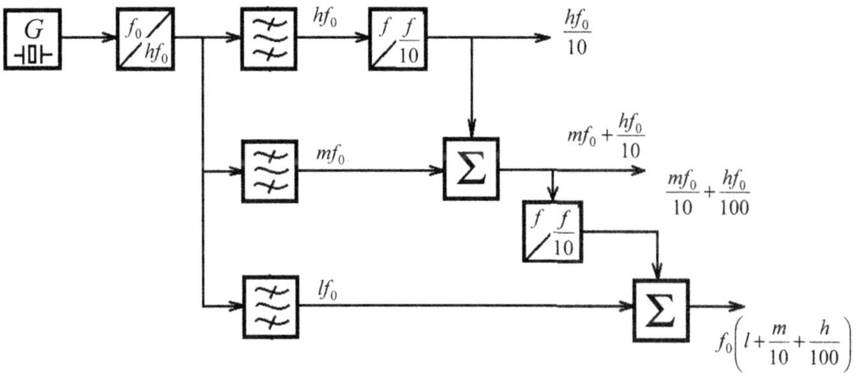

Fig. 5.12 The simplified block diagram of a decade-long synthesizer

highly stable reference generator generates a spectrum of ten adjacent frequency f_0 harmonics, from which one of the ten indicated harmonics hf_0 can be selected. The selected frequency value hf_0 is divided by ten and added to the selected other m-th harmonic (out of ten) of the frequency f_0. The result is the sum of the frequencies $mf_0 + \frac{hf_0}{10}$ s. This frequency value is again divided by 10 and added to some l-th harmonic of the reference frequency f_0.

$$f = f_0\left(k + \frac{l}{10} + \frac{m}{100} + \frac{h}{1000} + \ldots\right)$$

Coefficients *k, l, m* characterize the number of synthesizer steps. The choice of the number of steps is determined by the requirements for the minimum step of frequency tuning. So, if you choose $f_0 = 1$ Hz and provide six steps of tuning in the synthesizer, then the minimum step of discreteness will be 10 Hz.

The advantage of this scheme is mono-quartz frequency stabilization, which, with double temperature control, makes it possible to obtain the value of frequency instability $10^{-7} - 10^{-8}$.

In panoramic and measuring receivers, additional requirements are imposed on the narrowband path, which are due to the purpose of the RRR. In a sequential analysis receiver, the narrowband path is under the influence of voltage, the frequency of which changes over time, and therefore the requirements for the resolution and dynamic range of the receiver increase.

There are a number of methods for improving the basic performance of sequential analysis radios. They pursue the goal of overcoming the contradictions between the resolution and the speed of restructuring. Since, in sequential analysis, the resolution is determined by the dynamic characteristic of the narrowband path, the problem is reduced to narrowing the dynamic bandwidth while maintaining the tuning rate in the span [13].

One of the methods consists in the fact that the tuning of the radio receiver is carried out at a variable speed, namely, in the signal-free parts of the frequency range, it is carried out at a sufficiently high speed, and when a signal appears in the passband of a narrowband path, the tuning speed is automatically reduced to the optimal value, which provides minimum dynamic bandwidth. With this method of analysis and a low load of the frequency range, it is possible to significantly improve the resolution of the receiver without reducing the maximum allowable speed of its tuning.

Automatic gain control systems are widely used to improve the resolution of range radio receivers (which is especially important for measuring and panoramic radios) and to increase their dynamic range. It is known that to resolve two signals in frequency, it is necessary that the level of the "saddle" between the responses to two signals adjacent in frequency at the output of the path does not exceed the level of a certain threshold, and the level of the smaller of the two signals, naturally, is higher than the threshold. Let's explain this with the help of Fig. 5.13. Figure 5.13, a shows responses 1 and 2 at the output of a narrowband path from two signals of the same level. The resolution in this case is equal to Δf_G.

If the level of the second signal at the same frequency is higher than the first, then the response at the path output will also be greater, the "saddle" between the two responses 1 and 2 (Fig. 5.13b) will be higher than the threshold, and the signals in this case cannot be resolved. They can be resolved only for some $\Delta f_2 > \Delta f_1$ (i.e., at different signal levels, the receiver resolution deteriorates). As the analysis shows, with the usual dynamic ranges of signals at the input of the receiver (60–80 dB), the introduction of AGC gives an improvement in the resolution by two to three times. This gain is the greater, the larger the dynamic range of signals at the receiver input. In this case, the required degree of performance of the AGC system is determined by the basic parameters of the radio receiver (resolution and tuning rate) and is ensured by the appropriate choice of the time constants of the AGC filters. The structural diagram of the AGC system remains traditional [14].

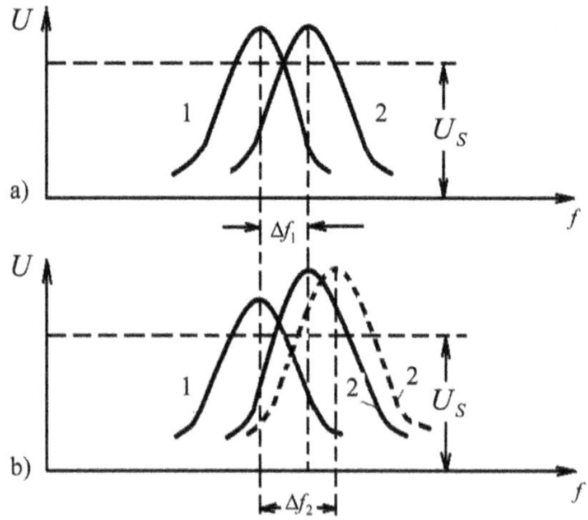

Fig. 5.13 Responses 1 and 2 at the output of a narrowband path from two signals of the same level

In some cases, logarithmic amplifiers can be used to equalize the signal level in a narrowband path. They are devices with a nonlinear (logarithmic) amplitude characteristic, in which there is a logarithmic dependence of voltage U_{out} on $U_{\text{in}p}$.

The gain of logarithmic amplifiers depends on the amplitude of the signal being amplified: it increases with decreasing signal amplitude and decreases with increasing signal amplitude.

The construction of many circuits of logarithmic amplifiers is based on two main methods: the method of nonlinear resistances and the method of sequential addition of voltages from the outputs of the amplifying stages. Nonlinear resistances are used either in the form of a load stage shunt or in a nonlinear feedback circuit in transistor or operational amplifiers. Note that the second method is used, as a rule, in the amplification paths of pulsed signals.

For example, a simplified circuit of a cascade with a load shunted by a nonlinear resistance R_1 is shown in Fig. 5.14, and with nonlinear feedback, in Fig. 5.15, where R_1 is a nonlinear resistance, the value of which varies depending on the level of the amplified signal. A junction in diodes or transistors can serve $p-n$ as a nonlinear resistance.

Fig. 5.14 A simplified circuit of a cascade with a load, shunted by a nonlinear resistance

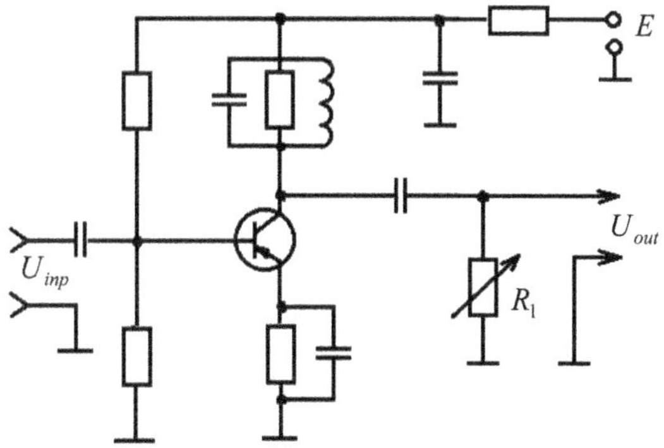

Fig. 5.15 A simplified circuit of a cascade with nonlinear feedback

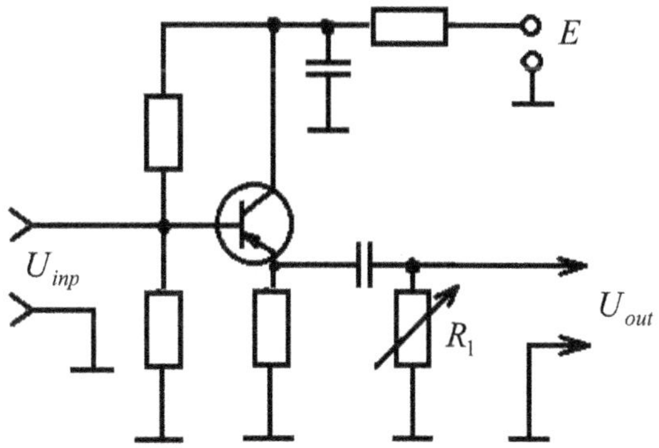

Indeed, the equation of the current–voltage characteristic $p - n$—the transition of a diode or transistor can be represented (if we neglect the saturation resistance) in the form

$$I_f = I_r(\exp(mU_f) - 1). \tag{5.5}$$

If we solve this equation for the forward bias voltage, we get a logarithmic dependence

$$U_f = \frac{1}{m}\ln\left(\frac{I_f}{I_r} + 1\right), \tag{5.6}$$

where I_f is the forward current of the diode or transistor;
U_f is the forward voltage of the diode or transistor;
I_r is the reverse saturation current.
The value is determined by the formula.

$$m = \frac{q}{\eta}kT, \tag{5.7}$$

where q is the electron charge;
k is the Boltzmann constant;
T is the temperature, K;
η is the diode (transistor) constant.

To obtain a logarithmic characteristic, a diode (transistor) used as a nonlinear element is fed with an initial bias voltage from a DC source. The sign voltage should be such that in the circuit shown in Fig. 5.14, the resistance R_1 decreases with increasing signal amplitude. Then the resulting full load resistance of the stage will also decrease, which will lead to a decrease in the gain.

The disadvantages of the considered amplifier include the inconstancy of the amplitude-frequency and phase-frequency characteristics of the amplifier due to a change in the equivalent load resistance.

In the circuit shown in Fig. 5.15, at low signal levels, the resistance R_1 should be small, and with an increase in the signal level, it should increase. Consequently, this should increase the degree of negative feedback, which will lead to a decrease in the amplifier gain. The nature of the change in the gain in this case will provide a logarithmic form of the amplitude characteristic of the stage.

The method of successive addition of voltages to obtain a logarithmic amplitude characteristic is illustrated in Fig. 5.16.

Here, the voltage from the outputs of several (N) series-connected IF amplifier stages with limitation at a certain level after detection is linearly summed by the adder. If the voltage at the amplifier input exceeds a certain potential, then the stage is saturated and it stops responding to an increase in the input signal level. The saturation mode first enters the last stage, and then the previous ones. In this case, as

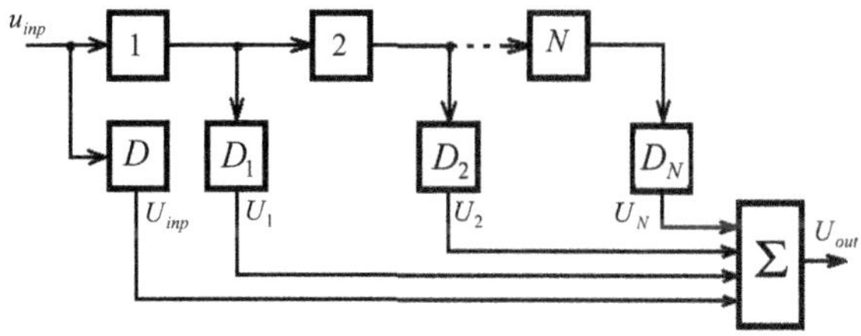

Fig. 5.16 The method of successive addition of voltages to obtain a logarithmic amplitude characteristic

a result of the addition of signals in the adder, it is possible to ensure the approximation of the amplitude characteristic to the logarithmic one [15].

Let's consider the process of obtaining the logarithmic characteristic for this circuit. In its most general form, the logarithmic amplitude characteristic is determined by the equation

$$U_{\text{out}} = a_1 \log(a_2 U_{\text{in}}). \tag{5.8}$$

or

$$U_{\text{out}} = a_1 \log(U_{\text{out}}) + a_1 \log a_2, \tag{5.9}$$

where a_1 is the slope of the logarithmic characteristic plotted on a logarithmic voltage scale; a_2—logarithmic shift, providing an increase in the output signal level by a constant value, as well as a shift U_{in} at $U_{\text{out}} = 0$, i.e.,

$$U_{\text{out}} = a_1 \log(a_2 U_{\text{in}}) = 0 \text{ or } U_{\text{in}} = \frac{1}{a_2}. \tag{5.10}$$

The view of the logarithmic characteristic is shown in Fig. 5.17. An amplitude characteristic close to a logarithmic one can be approximated by linear segments, the ends of which lie on an ideal characteristic.

Let there be a number of input voltages such as $U_{\text{in}1}, U_{\text{in}2}, U_{\text{in}3}, \ldots, U_{\text{in}N}$ that

$$\frac{U_{\text{in}N}}{U_{\text{in}N-1}} = \ldots = \frac{U_{\text{in}2}}{U_{\text{in}1}} = \frac{U_{\text{in}1}}{U_{\text{in}0}} = A,$$

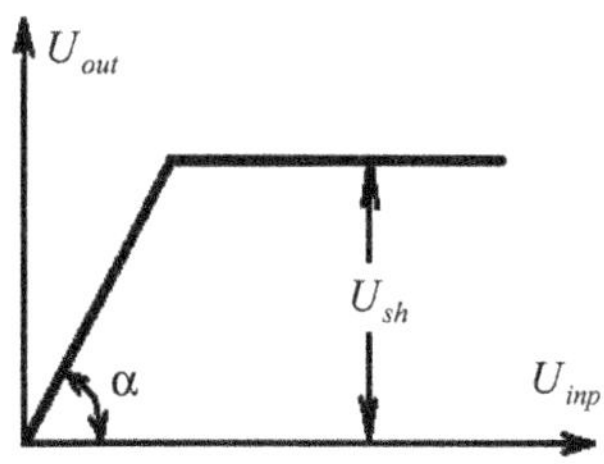

Fig. 5.17 The form of amplitude characteristic

that is, the ratio of adjacent values $U_{\text{in}\,i}$ and $U_{\text{in}\,i-1}$ is constant. Then we can write

$$U_{\text{in}\,N} = U_{\text{in}\,0} A^N. \tag{5.11}$$

Substituting Eq. (5.11) into (5.8), we obtain

$$U_{\text{out}\,N} = a_1 \log(a_2 U_{\text{in}\,0} A^N) = 0 \tag{5.12}$$

or

$$U_{\text{out}\,N} = a_1 \log(a_2 U_{\text{in}\,0}) + N a_1 \log A = 0, \tag{5.13}$$

as

$$U_{\text{out}\,0} = a_1 \log(a_2 U_{\text{in}\,0}), \tag{5.14}$$

then expression (5.14) is simplified and takes the form

$$U_{\text{out}\,N} = U_{\text{out}\,0} + N a_1 \log A = 0. \tag{5.15}$$

From the above it follows that the logarithmic functional dependence determines a number of input signals, the levels of which have the same ratio and which correspond to a number of output signals with levels having the same difference

$$\Delta = a_1 \log A. \tag{5.16}$$

We will assume that all stages (see Fig. 5.16) have an amplitude characteristic of the form shown in Fig. 5.17 U_b where is the limiting threshold, and $\text{tg}\,\alpha = K$ is the gain of the stage. Let's apply a sequence of voltages to the input of the amplifiers

$$U_1 = \frac{U_b}{K^N};\ U_2 = \frac{U_b}{K^{N-1}};\ \ldots;\ U_{\text{N-1}} = \frac{U_b}{K^2};\ U_N = \frac{U_b}{K}. \tag{5.17}$$

From here

$$\frac{U_{\text{N}}}{U_{\text{N - 1}}} = \frac{U_{\text{N - 1}}}{U_{\text{N - 2}}} = \ldots = \frac{U_2}{U_1} = \frac{U_1}{U_0} = K = A. \tag{5.18}$$

The total signal at the output of the circuit shown in Fig. 5.16 is

$$U_{\text{out}} = U_{\text{inp}\,0} + U_{\text{out1}} + +U_{\text{out}N}. \tag{5.19}$$

If we assume that all stages have the same gain and limiting levels, then the voltage at the output of the adder (provided that each is in limiting mode) can be represented as follows for different input signals. When a signal is applied to the input of the circuit, the last stage will enter the saturation mode $U_{\text{out }0} = \frac{U_{sh}}{K^N}$ first and the voltage at the output of the adder will be

$$U_{\text{out }N} = \frac{U_{sh}}{K^N} + \frac{U_{sh}}{K^{N-1}} + \ldots + \frac{U_{sh}}{K} + U_{sh}. \tag{5.20}$$

At the input voltage, the last and the penultimate stages will enter saturation and the voltage at the output of the adder will be

$$U_{\text{out }N-1} = \frac{U_{sh}}{K^{N-1}} + \frac{U_{sh}}{K^{N-2}} + \ldots + \frac{U_{sh}}{K} + 2U_{sh}. \tag{5.21}$$

By analogy, you can write

$$U_{\text{out }2} = \frac{U_{sh}}{K^2} + \frac{U_{sh}}{K} + (N-1)U_{sh}, \tag{5.22}$$

$$U_{\text{out }2} = \frac{U_{sh}}{K} + NU_{sh}. \tag{5.23}$$

If $K >> 1$, as usually happens in real circuits, then the terms of the sum to the left of the last terms can be neglected, i.e., the level of the total output voltage at the points corresponding to the transition to the limiting mode will receive a constant increment U_{sh} (with an increase in the input signal level by A times), which meets the requirements for the form of the amplitude characteristic that follow from relation (5.15). Thus, it is possible to form an amplitude characteristic that coincides with the logarithmic at discrete points.

Figure 5.18 shows a view of such a characteristic, where 1 is the true logarithmic characteristic and 2 is the amplitude characteristic formed by the method of successive addition. As you can see from the figure, there is a discrepancy in the ideal and real characteristics. The amount of error depends on the gain of the stages and the level of the input signals.

Note that the amplitude characteristic of any practically realized logarithmic amplifier at low input signal levels should differ from the ideal logarithmic characteristic, since $\log 0 = -\infty$. Therefore, usually for small signals, the resulting

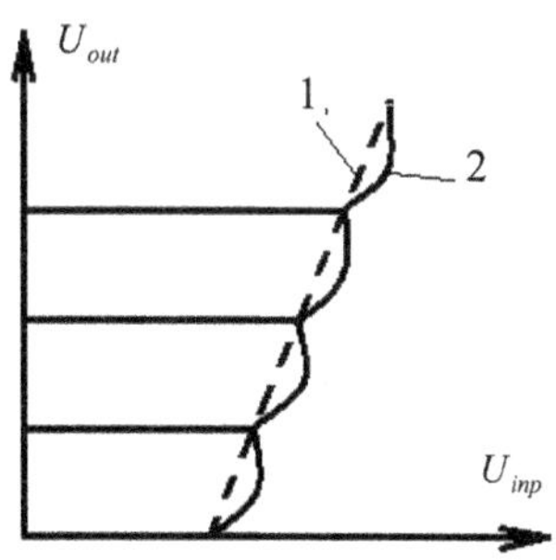

Fig. 5.18 A view of a characteristic

amplitude characteristic of the amplifier should be linear to the level of the amplifier entering the mode of logarithmic tracking of the input signal level.

5.5 Pairing the Settings of the Preselekgor and Heterodyne Circuits

In the range radio control systems, when they are tuned in any sub-range $f_{\min} - f_{\max}$ (Fig. 5.19), the conditions must be met

$$\begin{cases} f_G - f_{pr} = f_G - f_C = f_r \;\; for \;\; f_G > f_C; \\ f_{pr} - f_G = f_C - f_G = f_r \;\; for \;\; f_C > f_G, \end{cases} \quad (5.24)$$

where f_G, f_C, f_{pr} is the frequency of the local oscillator, signal and preselector settings, respectively.

When relations (5.24) are satisfied, i.e., when the RR is fine-tuned to the frequency of the received signal, its level at the output of the receiver will be maximum, since the frequency of the received signal is $f_C = f_{pr}$ and the signal is converted to an intermediate frequency without attenuation in the preselector. In a radio receiver with separate tuning of the local oscillator and preselector circuits, condition (5.24) is almost always fulfilled [16].

However, in radio receivers with one-hand tuning of the preselector and local oscillator circuits, i.e., restructuring by capacitors of variable capacity, structurally combined into one common block, in the entire range fails to ensure the fulfillment of relations (5.24). This is due to the fact that the minimum and maximum capacitances of each of the block's capacitors are, as a rule, the same, and therefore the overlapping coefficients of the frequency range $K_{OC} = \sqrt{\frac{C_{\max}}{C_{\min}}}$ will be the same when tuning the preselector and local oscillator circuits [10, 13]. However, in order to maintain a constant difference in tuning frequencies of these selective circuits, the coefficient of the preselector subband $K_{OC\,pr}$ and the coefficient of the local oscillator K_{DC} must be different. So, at the top tuning of the local oscillator

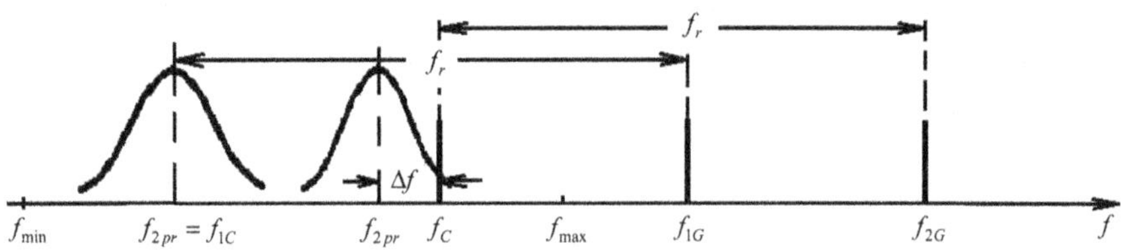

Fig. 5.19 The range radio control systems, when they are tuned in any sub-range $f_{\min} - f_{\max}$

$$K_{DC} = \frac{f_{G\ \max}}{f_{G\ \min}};\ K = \frac{f_{G\ \max} - f_r}{f_{G\ \min} - f_r}.$$

Consequently, this leads to the fact that when the receiver is tuned, some error Δf occurs in the preselector adjustment in relation to the signal, which causes its attenuation in the preselector. This attenuation will be the greater, the smaller the preselector passband and the squareness coefficient of its amplitude-frequency characteristic.

$$\Delta f = \frac{1}{2}\Delta f_{sh\ pr}, \tag{5.25}$$

where $\Delta f_{sh\ pr}$ is the preselector bandwidth.

In range receivers, to fulfill the condition (5.25), the so-called conjugation of the settings of the preselector and local oscillator circuits is carried out. There are several ways of conjugation, such as a phase shift between the plates of capacitors of individual capacitors of a common block of variable capacitors, the use of capacitors with different geometric shapes, etc.; however, these methods are often not technologically advanced. In practice, the conjugation method proposed by V.I. Siforov is most often used [17].

It consists in the use of additional coupling capacitors in the local oscillator circuit (at $f_G > f_C$) or in the preselector circuits (at $f_C > f_G$). The coupling method is based on the fact that two additional capacitors are included in the local oscillator circuit (at C_1 and C_2 (Fig. 5.20)), and the inductance is chosen such that equality is fulfilled in the middle of the range $f_G - f_{pr} = f_r$.

Let us explain the conjugation method using the graphs shown in Fig. 5.21, which shows the linear dependence of the change in the preselector tuning frequency from $f_{pr\ \min}$ $f_{\text{Прес min}}$ to $f_{pr\ \max}$ and the dotted line shows the case of ideal conjugation of the preselector and local oscillator frequency settings, in which the difference $f_G - f_{pr}$ over the entire tuning interval remains equal f_r. At point 0, the pairing is carried out due to the selection of inductance L_1, and at points 1 and 2 by the choice of capacitors C_1 and C_2. The values of the capacitances of the capacitors satisfy the inequalities $C_1 >> C_{\min}$ and $C_2 << C_{\max}$.

As a result, at the lower frequencies of the range at $C = C_{\max}$, the capacitance C_2 practically does not affect the capacitance of the loop, and the capacitance C_1, reducing the resulting capacitance of the loop, leads to an increase in the natural frequency of the loop. In this case, at point 1, the difference in the setting of

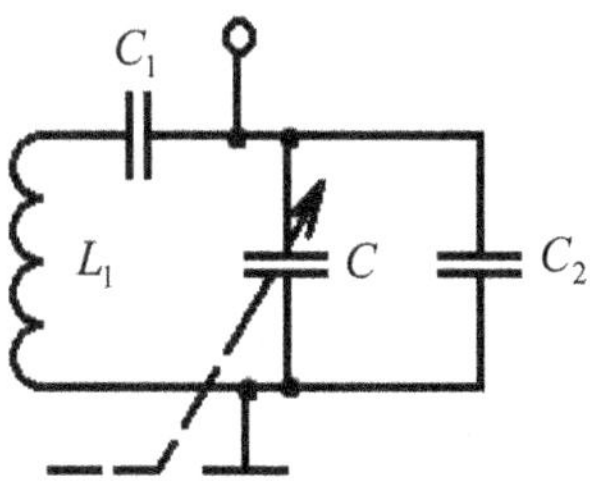

Fig. 5.20 The illustration of coupling method

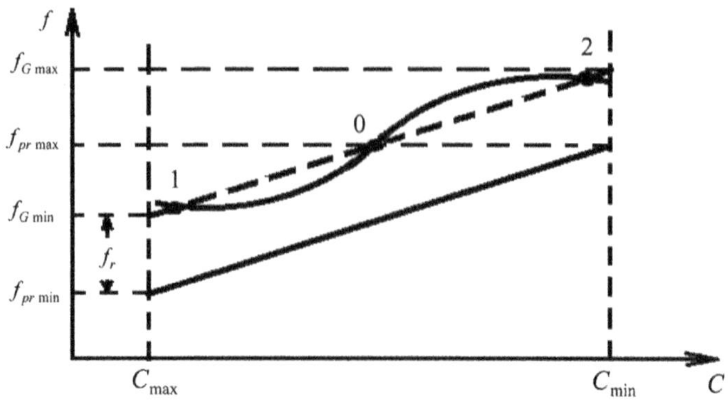

Fig. 5.21 The explanation of the conjugation method

the preselector and the local oscillator will exactly correspond to the intermediate frequency.

At high frequencies of the range $C = C_{\min}$, on the contrary, the capacitance C_1 practically does not play a role, and the capacitance C_2, increasing the resulting capacitance of the circuit, will lead to a decrease in the natural frequency of the local oscillator circuit, and at point 2 the circuit settings will be coupled. Thus, by selecting L_1, C_1, C_2 at three points of the subband, full conjugation of the settings will be ensured, and the change in the frequency of the local oscillator in the range of its tuning will be such that the condition $\Delta f < \Delta f_{pr}$ will be met in the entire range.

In the long-wave part of the radio frequency range, where the requirements for the matching of the loop settings, due to the small values of the coefficients of the subbands and the relatively wide preselectors, are not so high, accurate matching can be ensured at two or even one points.

5.6 Range Input Circuits and Amplifiers

In band radio receivers of decameter and longer wavelength ranges, as a rule, non-tuned antennas are used, i.e., antennas, the natural frequency of which, together with the coupling elements between the antenna and the input circuit, is outside the tuning range, and in shorter wavelengths, they are tuned. The amount of detuning is different for different types of antennas that can be used in the receiver. The parameters of antennas differ significantly from type to type and have great instability; therefore, in order to reduce the influence of antenna characteristics on the input circuit, its connection with the antenna is chosen weak [18]. Usually, with a weak connection from the antenna, an attenuation of no more than 20% of its own is introduced into the input circuit, which makes it possible to preserve the selective properties of the input circuit. In this case, the transfer coefficient of the input circuit turns out to be small.

In range radio receivers, single-loop input circuits with inductive and capacitive coupling with an antenna are most widely used. Complex electoral systems formed by several connected circuits are used in the input circuits in the event that increased selectivity requirements are imposed on the input device. It should be borne in mind that the increase in selectivity in such systems is achieved by reducing the gain, therefore, by increasing the noise figure and degrading the sensitivity of the RFA. By design, multi-circuit input devices are more complicated than single-circuit ones; therefore, in the range input circuits, the number of circuits used is small—no more than two or three. The ladder input circuits at the input of the receiver are seldom used [7, 14, 15]. They find application in receivers operating at one or more fixed frequencies, as well as in complex trunk receivers, the range of which is covered by a set of bandpass filters. When working at fixed frequencies, the filter elements are switched or a separate filter is connected at each frequency, and non-tunable filters can also be used, with a passband equal to the width of the frequency subband.

Let us consider several of the simplest circuits of range input circuits, by the example of which we will determine the dependence of the main characteristics of the input circuits on the frequency of their tuning.

An input circuit with inductive coupling with an antenna when the antenna is not tuned is characterized by the fact that in the impedance of the antenna circuit, taking into account the resistance of the coupling elements

$$z'_A = r_A + r_c + j(x_A + x_c) = r'_A + jx'_A, \tag{5.26}$$

the active resistance can be neglected in comparison with its reactance. Also, usually, taking these assumptions into account, the module of the impedance of the antenna circuit can be written as

$$z'_A = \omega L_1 - \frac{1}{\omega C_A} = \omega L_1 \left| 1 - \frac{\omega_A^2}{\omega^2} \right|, \tag{5.27}$$

where $\omega_A = \frac{1}{\sqrt{L_1 C_A}}$ is the natural frequency of the antenna circuit. In this case, the resonant transfer coefficient of the input circuit is equal to

$$K_{ic} = \frac{p_1 p_2}{z'_A C_E}. \tag{5.28}$$

Taking into account expression (5.27), the relations $p_1 = \frac{M}{L_2}$, $p_2 = 1$ for the incomplete switching coefficient for inductive coupling between the circuits, the relationship $K = \frac{M}{\sqrt{L_1 L_2}}$ between the two coils and the ratio

$$C_E = \frac{d_E}{\omega L_2}. \tag{5.29}$$

for the equivalent conductivity of the input circuit when it is rebuilt with a variable capacitance, one can obtain

$$K_{ic} = \frac{k}{d_E \left| 1 - \frac{f_A^2}{f^2} \right|} \sqrt{\frac{L_2}{L_1}}. \tag{5.30}$$

Let us make an approximate estimate of the dependence of the transmission coefficient of the input circuit on the frequency of its tuning and the natural frequency of the antenna circuit. In this case $d_E = \text{const}$, we assume that it is permissible for small values of the subband coefficient.

Three cases can be considered. If $f_{A1} >> f_{\max}$, then, in accordance with expression (5.30), the transmission coefficient of the input circuit is proportional to the square of the frequency to which the input circuit is tuned, physically this is explained by the fact that the antenna and the input circuit are two inductively coupled circuits tuned to different frequencies, and the EMF induced in the second loop under the influence of the antenna $\dot{I}_{A1}$ current is equal to

$$\dot{\varepsilon}_2 = -j\omega M \dot{I}_{A1}. \tag{5.31}$$

As the tuning frequency of the input circuit increases, this electro moving force (EMF) increases both due to an increase in the coupling resistance ωM equality and due to an increase in the current $\dot{I}_{A1}$ caused by the approach of the tuning frequency of the input circuit to the resonant frequency of the antenna (Fig. 5.22), which leads to an almost quadratic change in the value K_{ic} (curve 1) when rebuilding the input circuit. To prevent this impermanence from being extremely large is usually accepted $f_{A1} \geq (1,5-2) f_{\max}$.

An antenna whose natural frequency is higher than the maximum frequency of the receiver's sweep range is sometimes called a truncated antenna.

When $f_{A2} << f_{\min}$, as follows from expression (5.30), the transmission coefficient of the input circuit practically does not change in the range of its tuning (Fig. 5.22, curve 2). This is physically explained by the fact that the increase in the

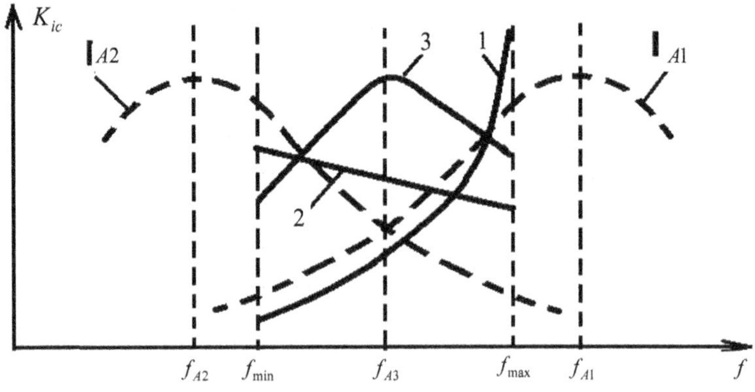

Fig. 5.22 The illustration of equality (5.30)

coupling resistance with increasing tuning frequency of the input circuit is compensated by the corresponding decrease in the current I_{A2} caused by the distance from the resonant frequency of the antenna. Usually, to exclude an unacceptable decrease K_{ic} when decreasing f_{A2}, it is taken $f_{A2} = (0,5 - 0,7) f_{\min}$. In this case, it decreases slightly as the tuning frequency of the input circuit increases. An antenna whose natural frequency is below the minimum frequency of the receiver's sweep range is sometimes called an extended antenna [4].

The third case, when $f_{\min} < f_{A3} < f_{\max}$, due to the extremely large dependence on the frequency of the quantity (Fig. 5.22, curve 3), is practically not used in the range input circuits.

To obtain the maximum transmission coefficient of the input circuit, the coupling coefficient between the antenna and the input circuit should be chosen in an optimal way, i.e., should be equal.

$$k_{0\ \text{opt}} = \frac{M}{\sqrt{L_1 L_2}} = p_{1\ \text{opt}} \sqrt{\frac{L_2}{L_1}}, \tag{5.32}$$

where $p_{1\ \text{opt}}$ is determined by relation (4.27). Note that when the input circuit is reconstructed, in addition to the change K_{ic}, its bandwidth also changes, which can be judged from relation (4.15), which, when the input circuit is reconstructed with a variable capacitance, can be written in the form

$$\Delta f_{ic} = f_0 d_E = \left[d + (p_1^2 G_1 + p_2^2 G_2)\omega_0 L\right]. \tag{5.33}$$

From relation (5.33), it follows that with an increase in the tuning frequency, the bandwidth of the input circuit increases sharply.

In the input circuit with external capacitive coupling with the antenna, the coupling capacity C_1 is small and since the antenna is not tuned, it can be neglected. Therefore, in the range of operating frequencies of the input circuit, the resistance of the antenna circuit has a capacitive character, and for an approximate assessment of the properties of the input circuit, we can consider

$$\frac{C_A + C_1}{\omega C_A C_1} >> r'_A \text{ and } \frac{C_A + C_1}{\omega C_A C_1} >> \omega L_A \tag{5.34}$$

on what basis

$$z'_A = x'_A = \frac{C_A + C_1}{\omega C_A C_1}. \tag{5.35}$$

Therefore, the transmission coefficient of the input circuit, taking into account (5.35) and (5.29), will be written in the form

$$K_{ic} = \frac{C_A C_1}{C_A + C_1} \frac{\omega^2 L}{d_E}. \tag{5.36}$$

Thus, when tuning the input circuit with a capacitance, the value K_{ic} increases with increasing tuning frequency in proportion to the square of this frequency. This is explained both by an increase in the current in the antenna circuit as the tuning frequency of the input circuit increases (because the antenna in this circuit is shortened due to its smallness C_1), and by an increase in the conductivity of the antenna coupling element with the input circuit. The bandwidth of the considered circuit, like the previous one, increases with an increase in the tuning frequency [17].

The high degree of variability of the transmission coefficient of the considered input circuit in the tuning range is its significant drawback. However, from a structural point of view, such an input circuit is simpler than an inductively coupled input circuit with an antenna.

To assess the nature of the change in the gain and bandwidth of band radio frequency amplifiers when tuning by their capacitance (which in practice prevails in the overwhelming majority of cases), we present them in the form:

$$K_0 = \frac{p_1 p_2 \left|\dot{Y}_{21}\right|}{p_1^2 G_1 + \frac{d}{\omega_0 L} + p_2^2 G_2}, \tag{5.37}$$

$$\Delta f_{sh} = f_0\left[d + (p_1^2 G_1 + p_2^2 G_2)\, 2\pi_0 L\right]. \tag{5.38}$$

If we assume that p_1 and p_2 do not depend on the operating frequency of the amplifier, then in the range of its tuning from relations (5.37) and (5.38) follows a very complex dependence of the main parameters of the cascade on the tuning frequency, since when the frequency changes, not only the parameters of the circuit change, but also parameters of electronic devices, which are taken into account in the last formulas through conductivity $\dot{Y}_{21}$, $G_1 G_2$,. In this case, the conductivity decreases $\left|\dot{Y}_{21}\right|$ with increasing frequency, while the conductivities G_1 and G_2 also increase, which usually leads to an increase in the value K_0, which manifests itself the more, the smaller the specific gravity in the equivalent load conductivity of the quantities G_1 and G_2.

Since the condition $K_0 \leq K_{0\,u}$ must always be met in amplifiers, the comparison of the resonant gain of the amplifier with the constant gain must be carried out at the upper tuning frequency of the amplifier.

The amplifier bandwidth, as follows from expression (5.38), sharply increases with increasing tuning frequency. Therefore, provided that the bandwidth of the receiver must be matched to the spectrum of the received signal, at the lower tuning frequency of the radio receiver, it should not be less than the permissible value from the point of view of possible distortions of the received signals, and at the upper frequency, it should be no more than the value at which the required receiver sensitivity occurs [15].

It should be noted that the conclusions that were made above did not take into account the frequency dependence of the coefficients p_1 and p_2 and their influence on the nature of the change in the amplifier parameters. So, for example, with transformer coupling of a circuit with an electronic device, depending on the natural frequency of the communication circuit, the same three cases of changing the amplifier gain in the operating frequency range are possible, which are obtained when considering an input circuit with inductive coupling with an antenna.

References

1. Voishvillo GV (1983) Amplifying devices. Radio and Communication, Moscow, p 264
2. Vorobiev VI (1983) Optical location for radio engineers. Radio and Communication, Moscow, p 176
3. Goldenberg LM, Matyushkin BD, Polyak MN (1985) Digital signal processing: a handbook. Radio and Communication, Moscow, p 312
4. Zernov NV, Karpov VG (1972) Theory of radio engineering circuits. Energiya, Leningrad, p 816
5. Vasilchenko NV, Borisov VA, Kremenchug LS, Levin GE (1983) In: Kurbatova LN, Vasilchenko NV (eds) Measurement of parameters of optical radiation receivers. Radio and Communication, Moscow, p 320
6. Kanevsky ZL (1963) Finkelstein M.I. Fluctuation interference and detection of pulsed radio signals. Gosenergoizdat, Moscow, Leningrad, p 243
7. McElroy JH (1977) Communication systems for near space using CO2 lasers. TIIER (2):54–89.
8. Malashin MS, Kaminsky RP, Borisov YB (1983) Basics of designing laser locating systems: textbook manual for universities. Higher shk, Moscow, p 207
9. Martynov VA, Selikhov YI (1980) In: Zavarin GD (ed) Pan-frame receivers and spectrum analyzers. Soviet Radio, Moscow, p 352
10. Pervachev SV (1982) Radioautomatics: textbook for universities. Radio and Communication, Moscow, p 296
11. Poberezhsky ES (1987) Digital radio receivers. Radio and Communication, Moscow, p 184
12. Pratt VK (1972) Laser communication systems: Translated from English. Svyaz, Moscow, p 232
13. Golubkov AP et al (1984) In: Sokolov MA (ed) Design of radar receiving devices: Textbook for radio engineering special universities. Higher School, Moscow, p 335
14. Teplyakov IM, Roshchin BV, Fomin AI, Weizel VA (1982) In: Teplyakova IM (ed) Radio systems for transmitting information: textbook for universities. Radio and Communication, Moscow, p 264
15. Ross M (1969) In: Nevsky AV (ed) Laser receivers: Translated from English. Mir, Moscow, p 520
16. Korostelev AA, Klyuev NF, Miller YA et al (1978) In: Dulevich VE (ed) Theoretical foundations of radar: Textbook for universities. Soviet Radio, Moscow, p 608
17. Xyuz RS (1980) Logarithmic amplifiers. Soviet Radio, Moscow, p 352
18. Vimberg GP, Vinogradov YV, Fomin AF et al (1972) In: Zenkiewic OA Energy characteristics of space radio lines. Soviet Radio, Moscow, p 436

Chapter 6
Receivers of Optical Systems

6.1 Characteristic of the Optical Wave Range

Optical waves are electromagnetic waves in the frequency range from 300 GHz to 3000 THz. In terms of wavelengths, the lower boundary corresponds to a wavelength in free space of 1 mm, and the upper one—0.1 μm (1000 Å).

The specified range is usually divided into three subranges corresponding to infrared (IR) radiation (wavelength from 1 mm to 0.75 microns), visible light (0.75–0.4 microns), and ultraviolet radiation (0.4–0, 1 μm).

Historically, the beginning of the development of the optical wavelength range is associated with the creation of passive systems designed to receive natural infrared radiation from various bodies, the temperature of which differs from absolute zero. Like radiometers, they are designed to detect objects, measure the distance to them, determine their temperature, etc. Active systems are being developed with the advent of optical quantum generators—lasers. Currently, the most intensively developed and widely used are laser communication systems and laser location systems (LLS), in the construction of which it is possible to realize almost all the advantages of optical systems over the corresponding radio engineering systems. Among these advantages, the following should be highlighted [1]:

(1) The possibility of allocating operating frequencies to a significantly larger number of simultaneously operating means with a sufficiently wide spectrum of signals, due to the large length of the optical wavelength range;
(2) The possibility of creating small in size, but very highly directional antennas, providing high secrecy of the corresponding systems and their immunity from deliberate interference;
(3) The possibility of increasing the accuracy of measuring the range, velocity, and angular coordinates of objects by an order of magnitude due to the use of ultra-short pulses and narrow antenna patterns in the implementation of laser ranging;

S. V. Dvornikov et al., *Radio-Electronic Equipment in Civil Aviation*, Springer Aerospace Technology, https://doi.org/10.1007/978-981-19-6199-1_6

(4) The possibility ensuring in laser communication lines the information transfer rate of the order of 1 Gbit/s or more, and in a multichannel line—of the order of 30 Gbit/s when using lasers generating short pulses with a high repetition rate;
(5) The possibility of receiving very weak optical signals due to the discrete registration method by the counting method of individual photoelectric of thrones.

However, the intensive development and implementation of optical systems in practice are constrained by the following drawbacks due to various reasons [2]:

(1) A relatively narrow mastered section of the optical wavelength range ($\lambda_c \approx 0.4 \div 10\ \mu m$) due to the lack of an appropriate element base for operation at other wavelengths;
(2) Strong attenuation of optical signals due to scattering in the atmosphere, especially in the presence of rain, snow and fog;
(3) Limited performance of modulators and demodulators of optical signals (the spectrum of reproducible modulation frequencies is limited from above by a value of about 10 GHz), which leads to the impossibility of realizing the huge potential bandwidth of the optical range;
(4) Inconsistency with modern requirements for mass, size, cost, efficiency and reliability of lasers (and the greatest opportunities are provided by coherent optical radiation);
(5) The complexity of searching for coherent optical signals by angular coordinates and frequency, as well as tracking a target from—for narrow, in particular, the directional patterns of the LLS antennas;
(6) The dependence of the accuracy of many location measurements on the distribution conditions straying of optical signals in the troposphere. It should also be noted that there is such an objective factor as the presence of additional noise in the optical range, which is due to the quantum nature of the optical signals themselves and leads to a limitation of the potentially achievable sensitivity of receiving devices.

6.2 Brief Information on Optical Radiation Receivers

Optical radiation receivers are designed to detect and measure the energy of electromagnetic waves in the optical range by converting it into other types of energy.

According to the principle of interaction of radiation with the substance of the sensitive element, optical radiation detectors, which are distinguished by a large variety, are divided into the following two large groups [3, 4]:

(1) Thermal radiation detectors (TPI);
(2) Photon (quantum) radiation detectors (FPI).

The action of TPI is based on a change in certain physical properties of the substance of the sensitive element due to the heating of the latter by radiation.

By the type of sensitive element, TPI can be divided into the following types of receivers:

- Bolometric radiation detectors, the action of which is based on a change in the complex resistance of a sensitive element made of metal, semiconductor or dielectric, when it absorbs incident radiation (which leads to a change in voltage in the corresponding circuit);
- Radiation thermoelements, in which the so-called thermoelectric effect is used, i.e., the effect of EMF on a thermocouple when heated by its radiation;
- Pyroelectric radiation detectors, the action of which is based on a change in the degree of spontaneous ionization of a pyroactive crystal (for example, a ferroelectric) with a change in its temperature;
- Optoacoustic (pneumatic) radiation receivers, the construction of which is based on the phenomenon of sounding gases in a closed vessel under the influence of an intermittent radiation flux.

One of the main disadvantages of TPIs is their increased inertia (with the exception of pyroelectric radiation detectors), as a result of which they can be used only when receiving relatively narrow-band signals. In addition, TPIs have low sensitivity over a wide spectral range (from 0.1 μm to hundreds of micrometers).

Thermal radiation detectors are currently used in instrumentation and radiometric devices [4]. In addition, the experience of using pyroelectric radiation detectors in the construction of coherent optical systems in the wavelength range of $\lambda_c \approx 10$ μm is known, but their sensitivity is still low.

In photonic (quantum) radiation detectors, direct interaction between radiation photons and electrons of the material of the sensitive element takes place. All of them are essentially quantum counters that respond to individual photons of the received radiation with the corresponding quantum efficiency η for a given wavelength, by which we mean the ratio [5]:

$$\eta = \frac{\overline{m}}{\overline{n}}, \tag{6.1}$$

where $\overline{m}$ is the average number of photoelectrons; $\overline{n}$ is the average number of quanta in the incident radiation.

The value η is determined by the properties of the material of the sensitive element and the wavelength of the optical signal and can range from 10^{-5} to units.

According to the principle of operation of the sensitive element, FPIs are divided into the following two main groups [2, 4]:

(1) Photoemission radiation detectors, the operation of which is based on the phenomenon of external photoelectric effect;
(2) Photoelectric semiconductor radiation detectors, the principle of operation of which is based on the phenomenon of internal photoelectric effect in semiconductors.

The spectral range of sensitivity of photoemission radiation detectors is determined by the characteristics of the light transmission of the vacuum shell window and the material of the photocathode and is in the range from 0.25 to 1.2 μm.

The most common device with an external photoelectric effect is photomultiplier tubes (PMTs), in which, due to secondary electron emission from dynodes, the flux of photoelectrons is significantly enhanced. The gain of modern photomultipliers reaches the value $10^6 - 3 \cdot 10^7$. At the same time, the quantum efficiency of PMT photocathodes is relatively low: $\eta = (4 \times 10^{-4} \div 0.1)$. At present, photomultipliers have been created, the cathodes of which have a quantum efficiency $\eta = (0.15 \div 0.2)$ in the range of visible light.

The inertia of a photomultiplier is characterized by values of the upper boundary modulation frequency of the order of tens and hundreds of megahertz, and for experimental samples—by values of 3–5 GHz.

In some cases, phototubes are used, but their sensitivity is lower than the photomultiplier tube sensitivity, since for the phototube to operate normally, a certain threshold photocurrent is required, formed into a continuous electron beam, while photomultipliers are capable of registering each independent photoelectron.

At present, PMTs are the most efficient detectors of radiation in the visible light and ultraviolet wavelengths [6].

The most common devices with an internal photoelectric effect are photoresistors, which use the effect of photoconductivity, and photodiodes, which use the photovoltaic effect.

In the near-IR region ($\lambda_c = 0.75 \div 3$ μm), photodiodes, especially silicon and germanium ones, are effective radiation detectors. The quantum efficiency of photodiodes is higher than the quantum efficiency of a PMT at comparable wavelengths. For example, for silicon photodiodes at $\lambda_c = 0.9$ μm $\eta \approx 0.9$, for germanium at $\lambda_c = 1.5$ μm $\eta \approx 0.5$.

However, photodiodes are inferior to PMTs in sensitivity, since the latter use internal amplification of the photocurrent. The so-called avalanche photodiodes, which are a solid-state analog of a photomultiplier, are promising. They use the mechanism of impact ionization in the region of the strong field of the reverse-biased transition. Therefore, under certain conditions, they can replace PMTs in the wavelength range of more than 1 μm. However, the multiplication of carriers in avalanche photodiodes causes additional noise due to fluctuations in the multiplication factor. There are data [6] on the use of reverse-biased photodiodes with a quantum efficiency $\eta > 0.25$, based on the Hg Cd Te compound, at a wavelength of 10.6 μm.

In the middle ($\lambda_c = 3 \div 5$ μm) and far ($\lambda_c > 5$ μm) infrared regions, the main type of radiation detector is photoresistors. They retain a high quantum efficiency $\eta = 0.5 \div 0.6$ at a wavelength of $\lambda_c = 10.6$ microns, but are very inertial. However, when cooled to temperatures of the order of 4–30 K, the inertia of the photoresistors is significantly reduced, the maximum modulation frequency can reach hundreds of megahertz. Such cooling also makes it possible to minimize the thermal noise of the photoresistor and, accordingly, increase its sensitivity. However, the operation of complex and bulky cooling systems is possible in this case only in stationary conditions.

Optical radiation receivers are one of the main elements of the receivers of optical systems and perform in the latter the functions of photodetectors and optical mixers.

6.3 Structural Diagrams of Receiving Devices of Optical Systems

The design of optical signal receivers is based on two main reception methods: direct photodetection and the superheterodyne method. These methods are inherently similar to those used in the radio frequency band.

Receivers with direct photodetection. The direct photodetection method is often called the energy method since a photodetector built on the basis of this method reacts only to the energy of the received optical signal and does not respond to changes in its phase and frequency. Therefore, the optical signal at the input of the photodetector in a receiver with direct photodetection must be modulated in amplitude (intensity), i.e., the signal must be amplitude or pulse modulated. Since, in this case, photonic (quantum) radiation detectors are used as photodetectors in the receivers of special systems and complexes, the quantum nature of the optical field manifests itself during detection [7].

Receivers with direct photodetection (power receivers) can operate in two modes:

(1) Photoelectron counting mode (discrete mode);
(2) Envelope playback mode (analog mode).

The photoelectron counting mode occurs when weak optical radiation is received, when a time-separated photon flux arrives at the photodetector. In this case, the signal envelope is not reproduced at the output of the receiver, and a certain number of single-photon current pulses, which do not overlap in time, fall on the allocated time interval, for example, the duration of the pulse signal τ_p. This mode of operation is typical for laser radar stations that locate distant objects, such as the Moon. The post-detector part of the receiving device can be constructed in a variety of ways. One of the options for constructing a receiver operating in the photoelectron counting mode is shown in Fig. 6.1. Hereinafter, the elements of the optical receiving antennas are not shown.

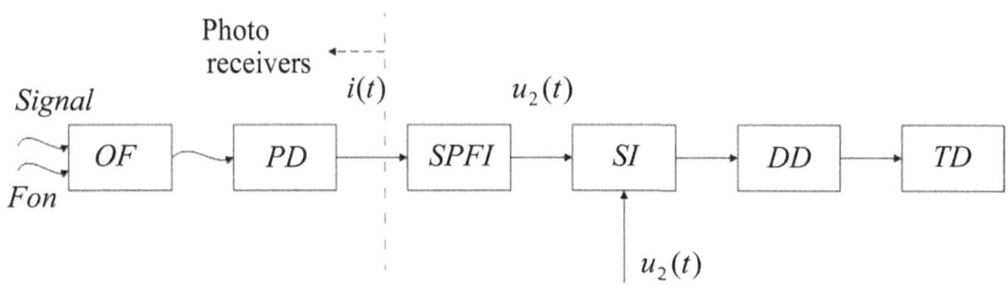

Fig. 6.1 The option for constructing a receiver operating in the photoelectron counting mode

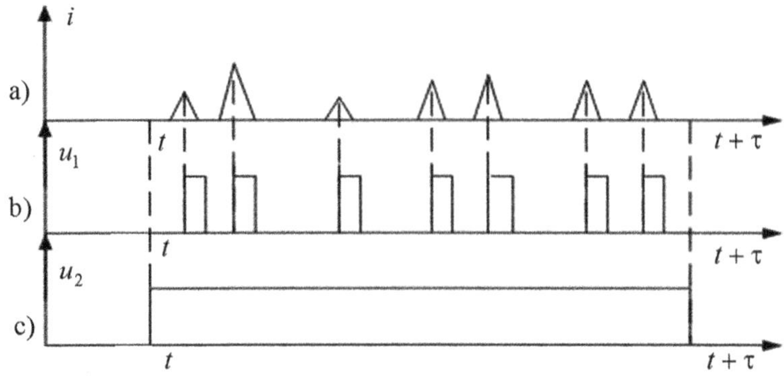

Fig. 6.2 The timing diagram of the gate signal

An optical filter (OF) is designed to suppress background radiation. Interference or polarizing filters, which have a fairly wide passband, of the order of 600–3000 GHz, can be used as the OF. The best 0F samples have a bandwidth of about 40 GHz.

As a photodetector (PD) in such circuits, photomultiplier tubes are most often used. The part of the receiving device from the entrance to the detector, inclusive, is usually called a photodetector. The timing diagram of current pulses in the load circuit of the photodetector is shown in Fig. 6.2a.

From the output of the photodetector, the signal enters the radio-technical part of the receiver, which in this case is built on logical (discrete) elements. The timing diagram of the output signal of the standard pulse former (SPF) is shown in Fig. 6.2b.

The output device of the receiver (see Fig. 6.1) is implemented in the form of a serial connection of a digital integrator, consisting of a counter of formed standard impulses (SI), a threshold device (TD) and a deciding device (DD) that decides on the presence of a signal at the output of the receiver when the threshold value entered in the control panel is exceeded, which is determined by the signal processing algorithm. The timing diagram of the gate signal $u_2(t)$ is shown in Fig. 6.2c.

Since in the process of receiving and processing the signal is understood as the average number of photoelectrons at the output of the photocathode of the PMT during the signal pulse duration, the photoelectron counting mode implements the matched filtering of the input signal in the receiver and provides the highest probability of correct reception [8].

The envelope playback mode (analog mode) is implemented at a high intensity of the optical signal, when the single-photon current pulses follow so densely that they merge into a single–signal envelope. At the same time, analog devices are used for registration, which greatly simplifies the post-detector processing circuit.

The block diagram of a power receiver operating in the envelope playback mode is shown in Fig. 6.3.

The output signal of the photodetector (Fig. 6.4a) through the low-pass filter (LPF), where it is "smooth" (Fig. 6.4b), enters the video path (VT).

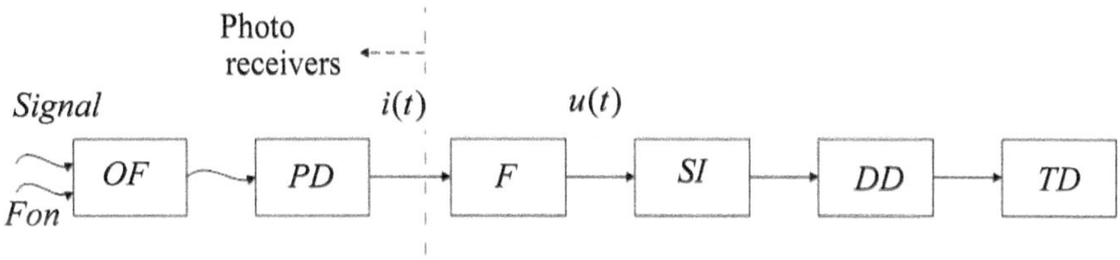

Fig. 6.3 The block diagram of a power receiver operating in the envelope playback mode

Fig. 6.4 The output signal of the photodetector through the low-pass filter

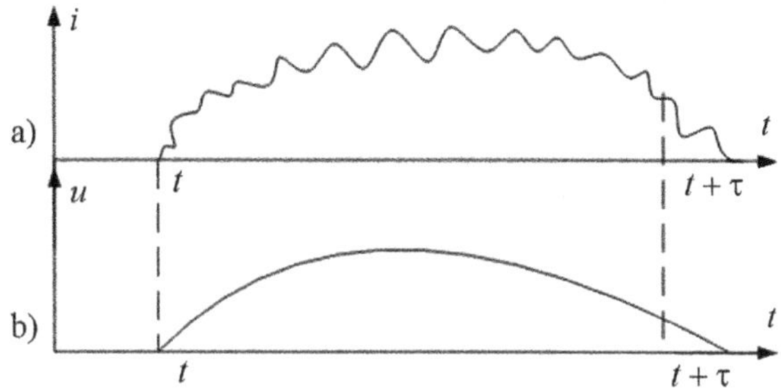

Optical signal detection is carried out in the same way as in radar. The deciding device (DD) makes a decision on the presence of a signal when the threshold device exceeds a certain level by the instantaneous value of the output voltage of the video path. This decision is made taking into account the optimal processing in the video path of either a single pulse or a burst of pulses. This design of the receiving device is typical for pulsed laser range finders and for LLS target tracking in the direction. If the transmission is carried out with subcarriers, which are selected, as a rule, in the microwave range, then after the photodetector, the subcarrier signal, modulated by the information message, enters the input of a conventional radio receiver, the circuit of which provides the required signal processing and conversion.

Superheterodyne receivers: The use of a heterodyne reception method in the optical wavelength range is possible only when using lasers with a small width of the radiation spectrum. Currently, superheterodyne receivers are used mainly at wavelengths of about 10.6 μm, since lasers CO_2 do not work well as stable local oscillators. The block diagram of a superheterodyne optical receiver is shown in Fig. 6.5.

In such a scheme, between the radiation of an optical heterodyne (OH), the role of which is played by a special laser, and the received optical signal, beats occur and the signal is converted into an intermediate (difference) frequency signal. The functions of an optical mixer (OPM) are performed by a semitransparent mirror (SM) and the mixer itself (photodetector). In this case, the optical signal is converted into a radio frequency signal, which provides effective filtering of the signal in conditions of external background radiation due to the high selectivity of the radio receiver [9, 10].

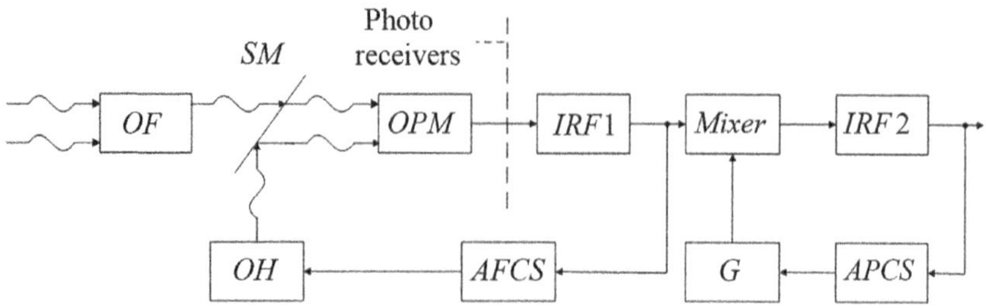

Fig. 6.5 The block diagram of a superheterodyne optical receiver

The main difficulty in the technical implementation of the scheme lies in the need to use an automatic frequency control system (AFCS) of an optical heterodyne with very large capture and hold bands. This is due to the fact that the Doppler frequency shift along the optical carrier can reach values of the order of several or even tens of gigahertz. Often, to compensate for the Doppler shift, appropriate frequency target designations are used. Frequency tuning capabilities of existing laser heterodyne do not allow full tracking of the Doppler frequency shift of the input signal. Therefore, the amplifier of the first intermediate radio frequency (IRF1) is made sufficiently broadband, the second frequency conversion (Cm, G) is used, and automatic frequency control of the second local oscillator (SLO) is carried out.

Currently, superheterodyne receivers are mainly used in laser communication systems and in laser meters of the radial velocity of objects, where they use modulation of the laser radiation of the transmitter by microwave oscillations (subcarrier frequency), and the radial velocity of the object is measured by the Doppler shift of the subcarrier of the received signal. Receivers with direct photodetection and superheterodyne type have specific advantages and disadvantages. Their analysis will be carried out after considering the noise ratios when receiving optical signals.

6.4 Sources of Noise in Optical Systems

Optical lines differ from radio lines not only in the size of the carrier frequency. In some cases, these quantitative differences lead to significant qualitative changes. For example, there is no thermal noise in the visible light range, which means that such elements of photodetectors as antennas, optical filters, etc., do not create thermal noise, despite active signal losses in these elements.

There are the following main sources of noise in receiving devices of the optical wavelength range [5, 11]:

- Thermal noise of the photodetector;
- Flicker-noise of the photodetector;
- Noise of the dark current of the photodetector;
- Noise caused by the surrounding background;
- Quantum noise.

The thermal noise of the photodetector is generated by the load of the photodetector and subsequent circuits. It is similar in nature to the thermal noise of radio receivers and is a Gaussian random process.

Flicker noise (flicker noise) is the low-frequency fluctuation of the photodetector current. For example, in photoemission radiation detectors, this effect is associated with a time variation of emission from relatively large regions of the photocathode surface. The specified type of noise is characterized by the fact that the spectral density of its power decreases with increasing frequency (this fact is associated with the sometimes used name I/f "noise"). The noise spectrum is actually limited from above by a frequency value of the order of 100–200 Hz. The effect of flicker noise can be reduced by rejecting low frequencies at the output of the photodetector, but the signal should not have significant components in the low-frequency region. The latter is achieved either by using subcarriers for modulation, or by using pulsed transmission methods, in which signal distortions in the low-frequency region can be neglected. In general, it can be noted that flicker noise is not a determining factor and can usually be ignored.

The noise of the dark current of the photodetector is the fluctuations of the current carriers arising in the photodetector in the absence of external irradiation. For example, in a photomultiplier, the dark current flows due to thermionic emission from the cathode (and on a somewhat smaller scale from the first dynode), due to current leakage and the so-called regenerative ionization, in other words, as a result of positively charged ions hitting the cathode or dynodes as well as photons from excited gas molecules and glass fluorescence in the region behind the cathode. For good photocathodes at an ambient temperature of T = 300 K, the dark current density is $j_{T.T.} = 10^{-16}$A/sm^2.

The noise of the dark current of the photodetector can be reduced by lowering the temperature of the photodetector, as well as by special technical methods.

Background noise is the statistical fluctuation of radiation from the surrounding space (background), such as the firmament and the surface of the Earth. Such noises are successfully dealt with by reducing the intensity of the background radiation incident on the photodetector, and this can be achieved in three ways [7, 10]:

1. Spectral—by installing an optical bandpass filter in front of the photodetector with the narrowest possible passband;
2. Spatial—by narrowing the "receiver's viewing angle" in the presence of extended background sources. This method has no analogy in radio systems and is associated with the specifics of the radiation patterns of optical systems;
3. Temporary—using the technique of short impulses, which will be discussed below.

Quantum noise is a determining factor in the optical wavelength range. Quantum noise, which is sometimes called photon or radiation noise, arises from fluctuations in the number of photons of radiation relative to its average value $\overline{n}$.

Let us consider the nature and characteristics of quantum noise using the example of a receiver with a photodetector operating on the principle of external photoelectric effect.

The optical signal at the input of the photodetector with power P_C is, according to quantum theory, a stream of photons with an average number of photons per second $\overline{n}_C$. Then:

$$\overline{n}_C = \frac{P_C}{hf_C}, \tag{6.2}$$

where h is Planck's constant $h = 6.625 \times 10^{-36}$ (W.c)/Hz; f_C—optical signal frequency, Hz.

Photons knock out photoelectrons from the photocathode surface with a certain quantum efficiency. By virtue of the uncertainty principle of quantum mechanics, photoelectrons appear at random times and the photoelectron flux is a random process. When receiving an unmodulated signal, the average number of photoelectrons over the observation interval should be identified with the signal, and fluctuations in the number of photoelectrons relative to the average value should be identified with noise, which is called quantum noise.

It should be noted the main features of quantum noise:

- Quantum noise is unavoidable, which follows from the uncertainty principle of quantum mechanics;
- Quantum noise occurs only in the presence of an optical signal.

When receiving pulsed signals with a duration τ_p, the amplitude of the electric pulse at the output of the photocathode should be understood as the value $\eta\overline{n}_C\tau_p$—the average number of photoelectrons per time τ_p, and the variance τ_p of the number of photoelectrons per time τ_p should be identified with the quantum noise power in the interval τ_p. It was shown in [11] that the photoelectron flux at the output of the photocathode can be considered Paussonian law for any weak signal, when the so-called signal energy per degree of freedom is much less than unity, i.e., when the condition is met:

$$\frac{\eta\overline{n}_C\tau_p}{\tau_p\Delta f_{PF}} \approx 1, \tag{6.3}$$

where Δf_{PF} is the passband of the optical filter in front of the photodetector.

Indeed, the quantity Δf_{PF}^{-1} determines the minimum duration of the signal at the output of the inertialess photodetector.

The quantity $\eta\overline{n}_C\tau_p$ is the number of current pulses (number of photoelectrons) with a duration of the order Δf_{PF}^{-1} of at the output of the inertialess photocathode in time τ_p. Then $\frac{\eta\overline{n}_C\tau_p}{\Delta f_{PF}}$ the product determines that part of the duration τ_p that is occupied by the output pulses. If this product is divided by τ_p, then you can get a value that is the reciprocal of the average duty cycle of the pulses at the output of the photocathode. Thus, the validity of the Poisson statistics means that the flux of photoelectrons at the output of the inertialess photocathode is rare. And this requirement is usually satisfied for all cases of practical interest [12].

According to Poisson's law, the probability of appearance in time τ_p is exactly K of photoelectrons defined by the expression

$$P(K,\ \tau_p) = \frac{(\eta \bar{n}_C \tau_p)^K}{K!} \exp(-\eta \bar{n}_C \tau_p). \tag{6.4}$$

The mean value of the Poisson probability density $\eta \bar{n}_C \tau_p$ is the amplitude of the signal pulse with the duration τ_p at the output of the photocathode, and the variance of the probability density, which is also equal $\eta \bar{n}_C \tau_p$ to the power of the quantum noise at the output of the photocathode. Then the signal-to-noise ratio in terms of the power at the output of the photocathode in the presence of only quantum noise will be

$$P_K = \left(\frac{P_S}{P_{noise}}\right)_{out} = \frac{(\eta \bar{n}_C \tau_p)^2}{\eta \bar{n}_C \tau_p}.$$

Assuming $\tau_p = \frac{1}{\Delta f_{sp}}$, where Δf_{sp} is the width of the optical pulse spectrum, taking into account expression (6.2), we have

$$\gamma_{P_K} = \frac{\eta P_S}{h f_C \Delta f_{sp}}. \tag{6.5}$$

According to expression (6.5), the product $h f_C \tau_p$ can be considered as the noise power at the output of the photocathode $P_{noise\ out}$. Therefore, when analyzing optical systems, it is conditionally assumed that the spectral power density of quantum noise is $-N_k = h f_C$ (conditionally, because the absolute value of the spectral density depends on the power of the received signal). In general, the expression for the spectral density of thermal and quantum noise is written as follows [12]:

$$N = \frac{h f_C}{\exp \frac{h f_C}{kT} - 1} + h f_C, \tag{6.6}$$

where k is the Boltzmann constant, $k = 1.379 \cdot 10^{-23}$ W/(Hz K); T is the ambient temperature, K.

Based on the ratio $h f_C / kT$, two characteristic areas of the pattern are distinguished:

(1) Wave, or classical (region of relatively low frequencies), at $h f_C / kT$, when the wave properties of radiation are manifested and N $\approx$ N_T = kT, i.e., noise is determined only by thermal radiation;
(2) Quantum (region of very high frequencies) at $h f_C / kT$, where the quantum properties of radiation are fully manifested and N $\approx$ N_T = hf_C, i.e., noise is determined by quantum effects.

It should be noted that the fluxes of photoelectrons at the output of the photocathode, caused by the background radiation and the dark current of the photodetector, are also Poisson.

Consider the action of the signal and background on a photodetector. At low energy of these emissions, the resulting flux of photoelectrons will be Poisson with an average number of photoelectrons per second

$$\eta \overline{n}_{\Sigma} = \frac{\eta\,(P_C + P_F)}{hf_C} = \eta\,(\overline{n}_C + \overline{n}_F), \tag{6.7}$$

where P_F, $\overline{n}_F$—power and average number of photons per second of background radiation; $\overline{n}_{\Sigma}$ is the average number of photons per second of total radiation.

Expression (6.7) shows that in the case under consideration, the background does not affect the value of the useful signal at the output of the photocathode, which is equal to $\eta\overline{n}_C$. Physically, this can be explained by the fact that with rare fluxes of signal and background photons, the probability of coincidence of photoelectrons from the background and the signal is quite small, therefore, the fluxes of photoelectrons from the background and the signal do not interact. This implies a very important conclusion: unlike radio detectors, photodetectors do not suppress weak signal strong. As a result, a system with direct photodetection is as immune to interference as a radio system with a heterodyne (synchronous) detector.

6.5 Noise Ratio in Receivers with Fine Photodetection

In the general case, the equivalent noise circuit of the photodetector (without taking into account flicker noise) can be depicted in the form shown in Fig. 6.6.

In this circuit, the current amplifier (CuA) has a current gain K_i, and R is the load resistance of the photodetector. The circuit is valid for an inertialess photodetector. If

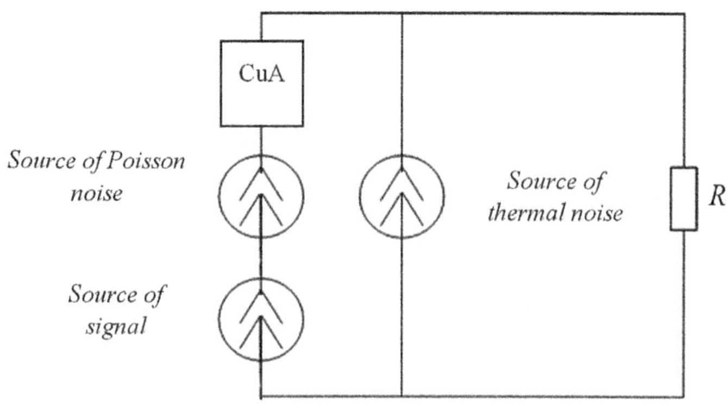

Fig. 6.6 The form of the photodetector (without taking into account flicker noise)

necessary, the inertia of the latter can be taken into account by including a low-pass filter at its output. As mentioned above, Poisson noise includes quantum noise, as well as noise caused by background radiation and the dark current of the photodetector. If the photoelectron flux is represented as a Poisson flux of δ pulses, each of which carries an elementary electric charge equal to the electron charge e, then the expression for the spectral density of Poisson noise N_P can be written as follows:

$$N_P = 2K_i^2 e^2 \overline{m} R, \tag{6.8}$$

where $\overline{m}$ is the average number of photoelectrons per second at the output of the photocathode [13].

The product $e\overline{m}$ is the average current at the output of the photocathode, which is the sum of the average current due to the presence of the signal $I_C = e\eta\overline{n}_C$, the average current due to the background $I_F = e\eta\overline{n}_F$, and the average dark current $I_{TT} = e\overline{m}_{TT}$, where $\overline{m}_{TT}$ is the average number of photoelectrons per second at the output of the photocathode in the absence of irradiation.

With the introduced notation, expression (6.8) takes the form

$$N_P = 2K_i^2 e(I_C + I_F + I_{TT}). \tag{6.9}$$

The resulting spectral density, taking into account the Poisson and thermal noise N, is determined according to the expression

$$N = N_P N_{noise} + N_T, \tag{6.10}$$

where N_{noise} is the factor that takes into account the increase in noise during the internal amplification in the photodetector (the so-called excess noise factor).

The value N_{noise} depends on the type of optical radiation receiver used as a photodetector.

So, for photomultipliers and avalanche photodiodes—$N_{noise} > 1$ for photodiodes with negative bias $N_{noise} = 1$, for photoresistors and photodiodes without bias N_{noise}.

Expression (6.10) taking into account (6.9) takes the form

$$N = 2K_i^2 e N_{noise}(I_C + I_F + I_{TT})R + kT. \tag{6.11}$$

Then the ratio of the signal and noise powers at the output of the photodetector in the video path band ΔF_P.

$$\gamma_P = \left(\frac{P_C}{P_{noise}}\right) = \frac{K_i^2 I_C^2 R}{N \Delta F_P} = \frac{K_i^2 \eta^2 \overline{n}_C e^2 R}{\left[2K_i^2 e N_{noise}(I_C + I_F + I_{TT})R + kT\right]\Delta F_P}. \tag{6.12}$$

After simple transformations, we get

$$\gamma_P = \frac{\eta \bar{n}_C}{\left[2N_{noise}\left(1 + \frac{I_F + I_{TT}}{I_C}\right) + \frac{kT}{eK_i^2 I_C R}\right]\Delta F_P}. \quad (6.13)$$

If all the noise at the output of the photodetector is negligible compared to quantum noise, then we get the maximum achievable signal-to-noise ratio (we assume $N_{noise} = 1$)

$$\gamma_P = \frac{\eta \bar{n}_C}{2\Delta F_P}. \quad (6.14)$$

Taking into account (6.2), expression (6.14) takes the form

$$\gamma_{P_K} = \frac{\eta P_C}{hf_C 2\Delta F_P}. \quad (6.15)$$

Expression (6.15) under the condition $2\Delta F_P = \Delta f_{sp}$ coincides with expression (6.5), obtained earlier based on the nature and statistical properties of quantum noise.

Consider the relationship between quantum and thermal noise in direct photodetection. Thermal noise power spectral density—$N_T = kT$. We find the spectral power density of quantum noise N_k using expression (6.9), but taking into account the excess noise ratio N_{noise}:

$$N_k = 2K_i^2 e I_C R N_{noise} = 2K_i^2 e \eta \bar{n}_C R N_{noise}. \quad (6.16)$$

The ratio of the spectral densities of thermal and quantum noise will be

$$\frac{N_T}{N_k} = \frac{kT}{2K_i^2 e \eta \bar{n}_C R N_{noise}} = \frac{kT}{2K_i^2 e \eta \bar{n}_C R N_{noise} \tau_P}\tau_P, \quad (6.17)$$

where τ_P is some duration of the information symbol.

Taking into account the value τ_P is necessary because in optical lines it is the product $\eta \bar{n}_C \tau_P$, that is, the average number of photoelectrons at the output of the photocathode over time τ_P, that determines the reliability of the information received. For real systems $\eta \bar{n}_C \tau_P = 10$, and the value τ_P is determined by the required information transfer rate in the system (for example, if the information transfer rate is 1 Mbit/s, then $\tau_P \approx 1$ μs).

Let us determine the requirements for the gain of the photodetector K at which the condition is satisfied $N_k \approx N_T$, i.e., receiver sensitivity is determined only by quantum noise.

Then $\eta \bar{n}_C \tau_P = 10$, in view of formula (6.17), we put the condition

$$K_i = \frac{1}{e}\sqrt{\frac{kT}{20RN_{noise}}\tau_P}. \quad (6.18)$$

Since for modern PMTs $T_i = 10^5 \div 10^7$, when they are used, the sensitivity of the receiver in many cases will be determined only by quantum noise. For example, at $T = 300$ K, $\tau_{и} = 1мкс$, $R = 100\ \Omega$, $Ш = 1$ from the formula (6.18) it follows that $K_i \gg 9000$. It should be noted that at a lower transmission rate, as can be seen from expression (6.18), a larger internal gain of the photodetector is required.

Let us now consider the previously mentioned method for reducing the influence on the reception quality of noise caused by background radiation and dark current of the photodetector, namely, a method based on the use of the short pulse technique [14].

Suppose that a pulse-code modulation-amplitude modulation (PCM-AM) modulated signal $u_C(t)$ with a chip duration is used in the laser communication line $\tau_{и}$ (Fig. 6.7). Let also the conditions be fulfilled under which the thermal noise can be neglected. Then the signal-to-noise ratio at the output of the photodetector (6.12) can be represented as

$$\gamma_P = \left(\frac{P_C}{P_{noise}}\right) = \frac{\eta^2 \overline{n}_C e}{2N_{noise}(I_C + I_F + I_{TT}\Delta F_P)}. \tag{6.19}$$

We define the values of the currents as $I_C = \eta\overline{n}_C e$, $I_F = \eta\overline{n}_F e$, $I_{TT} = \overline{m}_{TT} e$ Multiply and divide the expression (6.19) by the product $h^2 f_C \tau_P^2$ and take into account that the energy in the pulse is equal to $E_P = \overline{n}_C h f_C \tau_P$. Then expression (6.19) takes the form

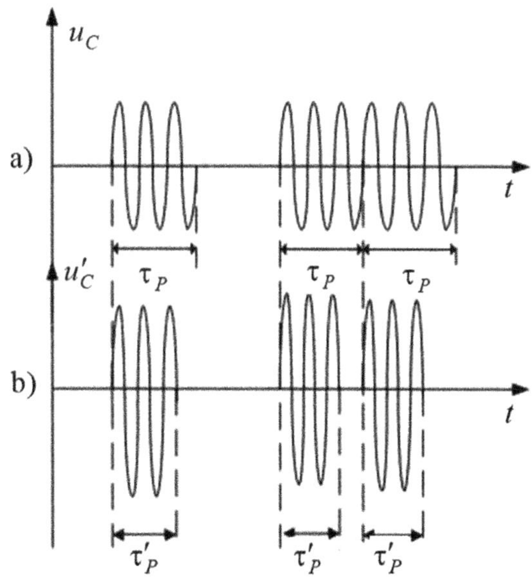

Fig. 6.7 The using of PCM-AM modulated signal with a chip duration in the laser communication line

$$\gamma_P = \frac{\eta^2 E_P^2}{2N_{noise}hf_C\tau_P\left[E_P + hf_C\tau_P\left(\overline{n}_F + \frac{\overline{m}_{TT}}{\eta}\right)\right]\Delta F_P}. \tag{6.20}$$

For a pulse signal, equality $2\Delta F_n \tau_P \approx f_{sp}\tau_P \approx 1$ is true. We finally get

$$\gamma_P = \frac{\eta E_P^2}{N_{noise}hf_C\left[E_P + hf_C\tau_P\left(\overline{n}_F + \frac{\overline{m}_{TT}}{\eta}\right)\right]}. \tag{6.21}$$

It can be seen that the signal-to-noise $E_P = \text{const}$ ratio with and decreasing the pulse duration τ_P (see the graph $u'_C(t)$ in Fig. 6.7) increases, i.e., the influence of the background radiation and the dark current of the photodetector on the reception quality is reduced. It is known that in the radio range, noise immunity when receiving discrete signals does not depend on the signal shape, but is determined only by the energy spent on symbol transmission, and in laser systems, noise immunity significantly depends on the shape (in particular, the range) of the pulse signals used.

6.6 Noise Rates in Superheterodyne Optical Receivers

Before considering the effect of noise on the quality of reception in superheterodyne optical receivers, let us turn to the energy relationships in the process of frequency conversion.

Let us denote the strength of the electric field of the signal on the surface of the photodetector acting as an optical mixer (see Fig. 6.5) through $e_C(t) = E_{m\,C}\cos\,\omega_C t$ similarly, for the heterodyne oscillation, we write $e_G(t) = E_{m\,G}\cos\,\omega_G t$. Here $E_{m\,C}$, $E_{m\,G}$ are the amplitudes of the electric field of the signal and heterodyne oscillations, respectively. We assume that full spatial phasing of the signal and the local oscillator is provided at the mixer input. The component of the square of the total electric field strength on the surface of the photodetector, which varies with the difference frequency $\omega_G - \omega_C$, is equal to $E_{m\,C}E_{m\,G}\cos(\,\omega_G - \omega_C)t$.

For the power of the corresponding optical signal incident on the photodetector, a similar expression can be written (power is proportional to the square of the electric field strength)

$$P(t) = 2\sqrt{P_C P_G}\cos(\,\omega_G - \omega_C)t. \tag{6.22}$$

This signal corresponds to a flux of photoelectrons with an average number of photoelectrons per second $\overline{m}(t) = \eta P(t)/(hf_C)$ and dispersion $\overline{m}(t)$ [10, 15].

The intermediate frequency current at the output of the photodetector will be equal (we assume that the photodetector has no internal amplification)

$$i_{IA}(t) = e\overline{m}(t) = \frac{2e\eta}{nf_C}\sqrt{P_C P_G}\cos(\omega_G - \omega_C)t =$$
$$= 2e\eta\frac{P_C}{nf_C}\sqrt{P_G/P_C}\cos(\omega_G - \omega_C)t = 2\sqrt{P_G/P_C}e\eta\overline{n}\cos(\omega_G - \omega_C)t =$$
$$= I_{m\,IA}\cos(\omega_G - \omega_C)t,$$

where $\overline{n}_C = P_C/(hf_C)$.

The effective value of the intermediate frequency current will be

$$I_{IA} = \frac{I_{m\,IA}}{\sqrt{2}} = 2\sqrt{P_G/P_C}e\eta\overline{n}. \tag{6.23}$$

As shown earlier, with direct photodetection, the current at the output of the photodetector (without internal amplification) is $I_C = e\eta\overline{n}_C$. Then, for a superheterodyne receiver, the quantity $G = 2\sqrt{P_G/P_C}$ can be considered as the internal gain of the photodetector during photomixing. Therefore, receiver thermal noise can be suppressed if this gain is large enough. This feature allows in many cases to significantly increase the sensitivity of the receiver and improve the signal-to-noise ratio. However, it should be remembered that with a large value of the ratio P_G/P_C, the noise caused by the radiation of the local oscillator can be decisive. Therefore, the magnitude of the local oscillator power must lie within strictly defined limits.

Let us determine the value of the signal-to-noise ratio at the output of the photodetector during photomixing in the passband of an intermediate frequency Δf_P amplifier for the case when all noise, except for the quantum one, can be neglected.

The power of the useful signal at the output of the photodetector, taking into account expression (6.23) and in the presence of internal amplification, is equal to

$$P_{IA} = K_i^2 I_{IA}^2 R = \frac{2P_G}{P_C}K_i^2(e\eta\overline{n}_C)^2 R. \tag{6.24}$$

By analogy with the input signal, we represent the power of heterodyne radiation in the form $P_G = h\overline{n}_G f_C$, where $\overline{n}_G$ is the average number of photons of heterodyne radiation per second.

Then expression (6.24) can be rewritten as follows:

$$P_{IA} = \frac{2\overline{n}_G hf_G}{\overline{n}_C hf_C}K_i^2 e^2\eta^2\overline{n}_C^2 R \approx 2K_i^2 e^2\eta^2\overline{n}_C\overline{n}_G R. \tag{6.25}$$

Equality (6.25) is true, since in practice we can assume $f_G \approx f_C$.

The power spectral density of quantum noise at the output of the photodetector is determined by expression (6.8), i.e.,

$$N_{KB} = 2K_i^2 e^2\overline{m}R \quad N_{KV} = 2K_i^2 e^2\overline{m}R. \tag{6.26}$$

The signal-to-noise ratio in the band Δf_P will be.

$$\gamma'_{P_{IA}} = \frac{P_{IA}}{N_k \Delta f_P} = \frac{2K_i^2 e^2 \eta^2 \overline{n}_C \overline{n}_G R}{2K_i^2 e^2 \eta^2 (\overline{n}_C + \overline{n}_G) R \Delta f_P} = \frac{\eta \overline{n}_C}{\Delta f_P} \frac{\overline{n}_G}{(\overline{n}_C + \overline{n}_G)}.$$

Since in most cases $P_G \approx P_C$, then $\frac{\overline{n}_G}{(\overline{n}_C + \overline{n}_G)} \approx 1$ and we finally get:

$$\gamma'_{P_{IA}} = \frac{\eta P_{IA}}{h f_C \Delta f_P}. \tag{6.27}$$

Let us determine the value of the signal-to-noise ratio in the bandwidth of the video path Δf_P of the heterodyne receiver [16]. We assume that a synchronous detector is switched on in the receiver after the intermediate frequency amplifier, which is an ideal multiplier of the input signal with the reference signal $U_0 \sin(\omega_G - \omega_C)t = U_0 \sin(\omega_0)t$. If there is an oscillation at the input of the synchronous detector $U_i \sin \omega_i t$, then at the output we get (taking into account the selective properties of the video path).

$$U_i \sin \omega_i t \; U_0 \sin \omega_0 t = 0.5 U_0 U_i \sin (\omega_0 - \omega_i) t.$$

If we put it $U_i \sin \omega_i t \; U_0 \sin \omega_0 t = 0.5 U_0 U_i \sin (\omega_0 - \omega_i) t \; 0,5 U_{O\Pi} = 10$, then it should be considered that the synchronous detector simply transfers the spectrum of the intermediate frequency oscillations to the video frequency region with a transmission coefficient equal to unity. If a signal pulse or sinusoidal oscillation $U \sin \omega_0 t$ and noise with spectral density N_K act at the input of the detector, then at the output of the synchronous detector the signal will have amplitude U and power at unit resistance U^2, and the spectral density of the noise will double and become equal $2N_K$, since the same beat frequency between the reference signal and noise are created by two noise components located to the left and to the right of the frequency of the reference signal at the same distance. Then, in the video band Δf_P, the signal-to-noise ratio is.

$$\gamma_{P_{IA}} = \frac{U^2}{2 N_K \Delta F_P}.$$

If the equality is satisfied $\Delta F_P = \frac{\Delta f_P}{2}$ and the signal-to-noise ratio in the intermediate frequency path is $\gamma_{P_{IA}} = \frac{(U^2/2)}{N_K \Delta F_P}$, then for the synchronous detector we obtain the well-known relation $\gamma_{P_{IA}} = 2\gamma'_{P_{IA}}$. Thus, taking into account expression (6.27), we have

$$\gamma_{P_{IA}} = \frac{2\eta P_C}{h f_C \Delta f_P} = \frac{\eta P_C}{h f_C \Delta F_P}. \tag{6.28}$$

A distinctive feature of superheterodyne optical receivers is the need to meet stringent requirements for the spatial phasing of wavefronts of signal and heterodyne oscillations.

6.7 Spatial Conditions for Effective Frequency Photo-Conversion

In the optical range, the signal wavelength turns out to be much smaller than the dimensions of the photosensitive region of the nonlinear element, on which the photo displacement is carried out. This leads to the fact that the spatial requirements for photo displacement are more stringent than for mixing in the radio range. To simplify the mathematical calculations, let us consider the operation of an optical mixer with a linearly extended photocathode. Let us assume that a signal luminous flux with an electric field $e_C = E_{m\ C} \sin \omega_C t$ strength is supplied to a photocathode of length l. Tired according to the element dx at the moment in time t, the electric field strength of the heterodyne luminous flux will have the form

$$e_G = E_{m\ C} \cos(\omega_G t - \theta_x), \tag{6.29}$$

where θ_x is the phase shift due to the retardation of the front of the heterodyne flow wave at the point x.

Since the spatial delay is equal in magnitude to the product $x \sin \phi$, where ϕ is the angle between the signal and local oscillator wavefronts, the phase shift can be represented as [17]

$$\theta_x = \frac{2\pi}{\lambda_G} x \sin \phi = \beta x, \tag{6.30}$$

where

$$\beta = \frac{2\pi}{\lambda_G} \sin \phi. \tag{6.31}$$

Taking into account that the characteristic of the mixer is quadratic (which is true in most practical cases), the current density from the site dx will be

$$\frac{di_x}{dx} = a(e_C + e_G)^2 = a\ [E_{m\ C}^2 \sin^2 \omega_C t + 2E_{m\ C} E_{m\ G} \sin \omega_C t \times$$
$$\times \cos(\omega_G t - \beta x) + E_{m\ G}^2 \cos^2(\omega_G t - \beta x)],$$

where a is the coefficient taking into account the photocathode sensitivity. Hence, the component of the current density of the difference (intermediate) frequency $\omega_{IA} = \omega_C - \omega_G$

$$\frac{di_{x\ IA}}{dx} = a(e_C + e_G)^2 = a\ E_{m\ C} E_{m\ G} \sin(\omega_{IA} t + \beta x).$$

By integrating the obtained expression over the area where photomixing is carried out, we find the value of the intermediate frequency current

$$i_{IA} = \int_{-i/2}^{i/2} a\, E_{m\,C} E_{m\,G} \sin(\omega_{IA} t + \beta x)\, dx$$
$$= \frac{2a}{\beta} E_{m\,C} E_{m\,G} \sin\left(\frac{\beta l}{2}\right) \sin \omega_{IA} t.$$

After simple transformations, we get

$$i_{IA} = al E_{m\,C} E_{m\,C} \frac{\sin(\beta l/2)}{\beta l/2} \sin(\omega_{IA} t). \tag{6.32}$$

Analysis of expression (6.32) shows that the intermediate frequency current is maximum at a value βl tending to zero, as $\frac{\sin(\beta l/2)}{\beta l/2}$ it tends to unity. This corresponds to the ideal spatial matching of the fronts of the signal and heterodyne light fluxes. When $\pi = \beta l/2$ there is no intermediate frequency current at the output of the photomixer. This, according to formula (6.31), corresponds to the equality

$$\sin \phi \approx \phi = \lambda_G / l. \tag{6.33}$$

Using expression (6.33), you can determine the requirements for the spatial phasing of the signal and heterodyne streams. So, for example, at the wavelength of the heterodyne signal $\lambda_G \approx 10\ \mu m$ and the linear size of the photocathode of the mixer $l = 1\ sm$ for effective frequency conversion, the angle must be at least an order of magnitude less than the value determined by formula (6.33), i.e., should be equal $\phi \approx 10^{-4}$.

6.8 Comparative Evaluation of Optical Receivers with Direct Photodetection and F0T0 Frequency Conversion

Each of the two main types of optical receivers has a number of distinct advantages. With respect to receivers with direct photodetection, these include [15]:

(1) Relative simplicity of implementation;
(2) Insensitivity to the Doppler shift of the signal frequency;
(3) The possibility of implementing the counting mode photoelectrons for receiving very weak light radiation, which is a time-separated flux of photons;
(4) The possibility of implementing more efficient methods of mutual pointing of transmitting and receiving antennas.

The only fundamental advantage of superheterodyne optical receivers is a more efficient suppression of background radiation. The latter is provided by both spectral and spatial signal selection.

Spectral filtering is carried out in a narrow-band path of the last intermediate frequency, which can be matched to the bandwidth of the received signal, while the bandwidth of optical filters that select the signal in receivers with direct photodetection, as a rule, is many times larger than the signal spectrum width.

Spatial selection is provided by the need for careful matching of the fronts of the heterodyne and signal light fluxes. Therefore, the background radiation, which does not coincide in direction with the signal, will be significantly attenuated during frequency conversion (see Sect. 6.7).

The requirements for spatial phasing of the signal and heterodyne streams impose severe restrictions on the construction of the optical part of the receiver. So, for example, the permissible dimensions of the input aperture of the receiving device (lens or parabolic mirror) are limited by the technical possibilities of creating an optical product in which phase dispersion does not yet play a significant role. When the difference in optical paths from different points of the collecting surface to the photocathode becomes comparable to the wavelength of the light wave, the spatial phasing of the signal and heterodyne fluxes is disturbed, which leads to a sharp decrease in conversion efficiency [9, 14].

In addition, it is sometimes argued that laser systems with heterodyning must meet more stringent mechanical tolerances for the fabrication of optical elements (for example, the surface roughness of antennas).

All of the above is true when it comes to direct photodetection and photomixing of the same light wavelength. However, in practice, systems with direct photodetection operate in the visible light range, while systems with heterodyning operate in the far IR range. At the same time, it is clear that with increasing wavelength, the criticality of meeting the requirements for optical elements decreases.

One of the practical problems of photomixing is the need for high temporal coherence of signal and heterodyne emissions, since otherwise beatings may occur, for example, between the radiation modes of the heterodyne, which will reduce the photomixing efficiency [12, 17].

It should be noted that the statement about the insensitivity of receivers with direct photodetection to distortions of the wavefront of the signal is valid only under the condition $d_K = d_0$ where d_K is the diameter of the sensitive surface of the photocathode and d_0 is the diameter of the received signal focused on it in the absence of distortions of its wavefront. To fulfill this condition, the receivers must have a wide field of view, which inevitably leads to an increase in the background illumination level. In this regard, the most promising is the use of diffraction-limited receivers of direct photodetection with adaptive correction of wavefront distortions.

And, finally, a comparison of expressions (6.28) and (6.15) shows that the signal-to-noise ratio at the output of a superheterodyne receiver is twice that at the output of a receiver with direct photodetection. However, it is imperative to compare systems in terms of noise immunity, taking into account specific modulation and coding methods. This is evidenced by at least the fact that when pulsed signals are used in a receiver with direct photodetection, there is no quantum noise in the pauses between pulses, and during photomixing it is constantly present, since it is practically determined by the radiation of the heterodyne.

References

1. Belousov AP (1959) Calculation of the noise figure of radio receivers. M: State publishing house of defense. prom-sti, p 135.
2. Van der Ziel (1973) Noise (sources, description, measurements): Translated from English. M.: Sov. radio, p 178.
3. Voishvillo GV (1983) Amplifying devices. M.: Radio and communication, p 264.
4. Vorobiev VI (1983) Optical location for radio engineers. M.: Radio and communication, p 176.
5. Goldenberg LM, Matyushkin BD, Polyak MN (1985) Digital signal processing: a handbook. M.: Radio and communication, p 312.
6. Ditkin VA, Prudnikov AP (1965) A guide to operational calculus. M.: Higher. shk., p 466.
7. Evtyanov SI (1948) Transient processes in receiving-amplifying circuits. M .: Radio and communication, p 210.
8. Vasilchenko NV, Borisov VA, Kremenchug LS, Levin GE (1983) In: Kurbatova LN, Vasilchenko NV (eds) Measurement of parameters of optical radiation receivers. M.: Radio and communication, p 320.
9. Pestryakov VB, Kuzenkov VD (1985) Radio engineering systems. M.: Radio and communication, p 376.
10. Prat VK (1972) Laser communication systems: Translated from English. M.: Svyaz, p 232.
11. A.P. Golubkov, et al (1984) In: Sokolov MA (ed) Design of radar receiving devices: Textbook for radio engineering special universities. M.: Higher school, p 335.
12. Bankov VN, Barulin LG, Zhodzishsky MI, et al. (1984) In: Barulina LG (ed) Radio receivers. M.: Radio and communication, p 272.
13. Zyuko AG, Korobov YF, Levitan GI, Simontov IM, Falco AI (1975) In: Zyuko AG (ed) Radio receivers. M.: Communication, p 400.
14. Teplyakov IM, Roshchin BV, Fomin AI, Weizel VA (1982) In: Teplyakova IM (ed) Radio systems for transmitting information: Textbook for universities. M.: Radio and communication, p 264.
15. Ross M (1969) In: Nevsky AV (ed) Laser receivers: Translated from English. Moscow: Mir, p 520.
16. Zhodzishsky MI, Sila-Novitsky SY, Prasolov VA, et al. (1980) In: Zhodzishsky MI (ed) Digital phase synchronization systems. M.: Soviet radio, p 208.
17. Sheremetyev AG (1971) Statistical theory of laser communication. Moscow: Communication, p 264.

Chapter 7
Measurement of Basic Characteristics of Radio Receivers

7.1 General Radio Test Information

Control of the parameters of radio equipment is carried out at the stages of its design, production, and operation. The purpose of proof tests is to verify that the main characteristics of the equipment meet the specifications. The values of these characteristics and the tolerances for their deviations from the nominal values are determined by the technical requirements for a particular product. The technical requirements themselves are determined by state standards, which are subdivided into the following standards: technical specifications (TS), acceptance rules, operation and repair rules, test methods, typical technological processes, etc. Of these, in particular, the technical specification standards determine the technical requirements for products during their manufacture, delivery and operation, including operational characteristics, acceptance rules, etc. Standards of technical requirements (TR) determine the requirements for the reliability and durability of products, warranty periods of work, etc. Test method standards establish procedures for sampling products for testing and test methods [1].

The composition of the radio receiving device (RRD) test is determined for a specific equipment in the TS and, in general, includes the following main types of tests: mechanical, climatic, electrical and radio engineering.

Radio engineering tests consist in measuring the corresponding characteristics of the equipment, a list of which is given in the TS. For radio receiving equipment, such characteristics primarily include the sensitivity and selectivity of the radio receiver, signal distortion, and the efficiency of regulation systems. In addition to the listed characteristics, other characteristics determined by the intended purpose of the receiver can be measured during testing. Below is a general methodology for conducting basic radio engineering tests of radio receivers for various purposes, both during their production and during maintenance, which can be implemented in both automated and non-automated control systems.

S. V. Dvornikov et al., *Radio-Electronic Equipment in Civil Aviation*, Springer Aerospace Technology, https://doi.org/10.1007/978-981-19-6199-1_7

7.2 Measuring the Sensitivity of Radio Receivers

The technique and features of measuring the sensitivity of radio receivers are determined in each specific case by the technical conditions at the radio receiver and depend on the type of modulation of the received signals, the range of operating frequencies and the purpose of the radio receiver.

The real and ultimate sensitivity, the sensitivity limited by noise and gain, are measured in the RRD.

The real sensitivity of the radio receiver is determined by the minimum signal level at the input of the radio receiver, at which the required output voltage or signal power is provided at a given ratio of the signal power to the interference power at the receiver output.

The limiting sensitivity usually characterizes the linear channel of the RRD and is determined by the minimum signal level at the input of the RRD, at which the ratio of the signal power to the power of the internal noise at the output of the linear part of the RRD is equal to unity. Most often, the concept of limiting sensitivity is used in relation to microwave receivers [2].

The sensitivity of the radio receiver, limited by interference, is determined by the minimum level of the input signal, at which the specified ratio between the nominal output power (voltage) and the power (voltage) of interference from various paths and nodes of the radio transmitter itself is ensured.

The sensitivity of the RPU, limited by the gain, is determined by the minimum level of the input signal required to obtain the rated output power (voltage) of the RRD.

In practice, the measurement of the sensitivity of the RRD is carried out, which most fully characterizes its functional purpose. So, for example, in broadcasting radio transmitters the real sensitivity is measured, in television radio transmitters of a black and white image for the image path, the sensitivity is measured, limited by gain, interference, synchronization, etc.

It should be noted that the sensitivity of the range RRDs is measured at least at two points of each sub-range [3].

The block diagram for measuring the real sensitivity of radio receivers is shown in Fig. 7.1.

Here, SSG is a standard signal generator that provides the formation of signals with a given type and modulation parameters, MD is a measuring device, AE is an antenna equivalent, which allows you to obtain voltages and currents at the receiver input, commensurate with the voltages and currents arising under the influence of a signal with a certain strength on a real antenna field, and excludes the influence of the SSG on the characteristics of the INPUT circuit. When measuring the sensitivity

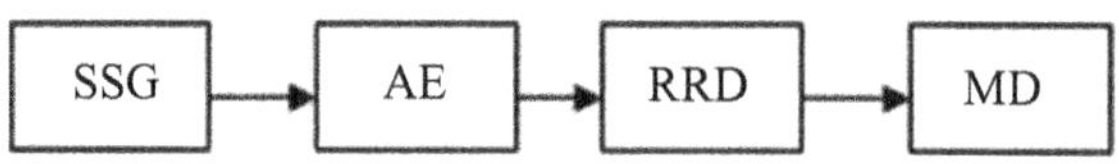

Fig. 7.1 The block diagram for measuring the real sensitivity of radio receivers

of receivers of the decimeter and shorter wavelength range in the structural diagram (Fig. 7.1), the equivalent of the antenna is usually absent. In this case, the output impedance of the SSG is used as the equivalent of the antenna. Therefore, when measuring the sensitivity of almost all receivers in this range, a SSG with a favorable conductivity equal to the equivalent conductivity of the corresponding antenna is used. This circumstance is explained by the fact that such receivers, as a rule, have a matched input. The figure also shows a measuring device (MD), which can be a voltmeter, ammeter, oscilloscope, spectrum analyzer, etc.

The measurement of the real sensitivity of the ASK receivers is performed at the set frequency of the radio control setup, the bandwidth and at the given parameters of the signal modulation, specified in the technical specifications. The AGC system (automatic gain control system) during measurements should be turned off, if possible. After tuning the SSG to the frequency of the receiver, select a signal level from the SSG at the input of the RRD, at which the output voltage required in the technical condition (TC) is provided at its output. Then, with the modulation turned off, by adjusting the gain of the IF amplifier (intermediate frequency amplifier) path, the required noise level at the output of the RRD is set. Repeating the measurement process with the signal modulation on and off, by means of gradual approximations, the required ratio of the signal level to the noise level at the receiver output is achieved. The value of the input voltage at which this condition is met is equal to the real sensitivity of the radio receiver.

The real sensitivity of the receivers of amplitude-shift keyed signals is measured according to the method described above. After the noise level permissible according to the TC is set at the receiver output with the SSG turned off, the SSG is turned on and an unmodulated voltage is supplied from it to the receiver input of such a level at which a normal output voltage is obtained at the output of the receiver [4]. The level of this signal determines the sensitivity of the receiver of the amplitude-shift keyed signals. In those cases when the shape of the output pulses is important, the real sensitivity can be determined by the minimum input signal level, at which the degree of pulse distortion at the RRD output does not exceed the permissible value.

The measurement of the real sensitivity of the receivers of frequency-modulated (keyed) and phase-shift keyed signals is carried out similarly to that discussed above, but with the AFC (automatic frequency control) system off, if any.

Sometimes, in receivers of frequency-shift keyed signals, the sensitivity is taken to be that value of the unmodulated input signal at which the output (shaping) stages from the receiver's own noise stop triggering, which also indicates a certain, required ratio of the signal level and interference at the output of the receiver [2, 5].

The sensitivity of receivers of television signals is measured separately for the image and sound channels. In this case, the image channel is usually estimated by sensitivity, limited by gain and interference. When measuring gain-limited sensitivity, set the maximum gain on the television receiver. An amplitude modulated signal with a frequency of 400 or 1000 Hz and a modulation depth of 50% is fed to the TV set input from the SSG. (If the receiver uses the AGC key circuit, then the voltage of the carrier frequency of the image is applied to the input of the receiver, modulated by a television video signal). The SSG is adjusted to the maximum of the output signal of the receiver and the input signal level is set at which the value of the output signal will be equal to the specified one. To determine the sensitivity limited by interference, a modulated voltage of the carrier frequency of the image, equal to the sensitivity limited by the gain, is supplied to the input of the receiver from the SSG. In this case, the receiver gain should be maximum. Next, the constant voltage in the AGC control circuit is measured. Then the modulation ν is turned off from an external source, a voltage equal to the previously measured voltage is supplied to the AGC circuit. The latter is necessary to maintain a constant mode of operation of the stages in the absence of modulation. When the modulation is off, the noise level is measured. If the ratio of the normal output signal to the noise level differs from the required one, then by successively changing the gain (contrast) and the input signal level, the required value of this ratio is achieved. In this case, the magnitude of the input signal will determine the sensitivity limited by interference.

The real sensitivity of the audio channel depends not only on the level of its own noise, but also on the level of interference created by the image path. Therefore, to measure the sensitivity, voltage is simultaneously applied to the input of the television receiver from two generators—the FSK (frequency-shift keying) and ASK generators [6].

The ASK generator is tuned to the carrier image frequency of the corresponding channel and modulated with a television video signal or voltage with a frequency of 1000 Hz with a modulation depth of 50%. A signal with an audio frequency simulated by a frequency of 400 or 1000 Hz with a deviation of 50 kHz is supplied from the FSK generator. The ratio of the input levels of the image signals and the audio signal must correspond to the specifications (usually the level of the audio signals is taken two times less than the level of the image signals). During measurements, this ratio must be kept constant. After fine tuning and obtaining normal output signals, the modulation of the SSG frequency is turned off and the level of the total interference of both channels is measured at the output of the audio channel. Then, keeping the ratio of the voltages supplied from the SSG constant, the output voltage of the generators is changed until the specified ratio of the audio signal level to the noise level is provided at the receiver output. The resulting value of the voltage supplied from the SSG FSK to the input of the receiver will be the real sensitivity in the sound channel.

The measurement of the real sensitivity of microwave receivers intended for the reception of signals with continuous radiation is performed in the same way as the measurement of the sensitivity of the receivers considered earlier. First, the AGC and AFC systems are turned off, and the intermediate frequency gain control is used to set the required noise level at the receiver output according to the technical specifications. Then from the SSG to the input of the receiver, a signal is supplied with the modulation parameters specified in the TC and the value of the input signal is set at which the ratio of the signal level to the level of the internal noise of the receiver at its output, required in the TC, is ensured.

The receiver sensitivity (RS) limit can be determined either by directly measuring it with the SSG or by calculating it through the noise figure (NF) measured with a noise figure meter. The calculation is carried out according to the formula

$$P_{RS} = NF\Delta f_{noise} \tag{7.1}$$

Measurement of the limiting sensitivity with the help of the SSG consists in determining the signal power at the receiver input, at which the total power of the signal and internal noise at the output of the receiver's linear path is twice the power of only the internal noise. The measurement sequence is as follows. When the SSG is off, the level of intrinsic noise specified in the TC is set at the output of the linear path of the RRD. Then, a modulated signal is fed from the SSG to the input of the RRD and its level is selected so that the output power of the linear path is doubled. In this case, the level of the input signal will correspond to the limiting sensitivity of the RRD.

The technique for measuring the sensitivity of receivers in the optical wavelength range largely depends on their types, which, as you know, are very diverse, which is associated with both the significant width of the optical range and the variety of methods for converting the energy of electromagnetic waves. The latter leads to the fact that practically each type of receiver of the optical receiving system introduces its own definitions of sensitivity. So, for example, for thermal detectors, current or volt sensitivity, etc. are measured, for photoemission ones, the light sensitivity of the photocathode, light anode sensitivity, spectral anode sensitivity, etc. All this leads to a significant variety of measurement techniques, and in each individual case, the technique is selected specifically for each optical receiver [7].

7.3 Measurement of Election

In radio receivers, the selectivity is measured by single-signal and multi-signal methods.

The single-signal selectivity method is used to assess the selectivity of a radio receiver operating in a linear mode. This method is used to remove the amplitude-frequency characteristic of the radio receiver, the characteristic of one-signal selectivity, measure the attenuation of interference in the adjacent channel and the degree of suppression of the side receiving channels.

The method of multi-signal (two-signal) selectivity is used to assess the selective properties of a radio receiver taking into account nonlinear processes caused by the simultaneous action of a useful signal and radio interference on its input. This method evaluates the effects of blocking, crosstalk and intermodulation [3, 5].

7.3.1 Single Signal Selectivity

The amplitude-frequency characteristic (AFC), which characterizes the selectivity of the main receiving path of the RRD, is taken either using a device for studying the AFC, or according to the diagram shown in Fig. 7.1, with a measuring device connected to the output of the linear path. In the first case, the measurement procedure is determined by the operating instructions for the device for studying the frequency response, and in the second it can be as follows. A modulated oscillation equal to the sensitivity of the receiver is fed to the input of the RRD, and (with the AFC system turned off), frustrating the SSG in both directions relative to the receiver tuning frequency, the dependence of the output voltage on the frequency detuning value is removed.

When taking the characteristics of single-signal selectivity in ASK receivers, a modulated oscillation equal to the sensitivity of the receiver is fed to the input of the RRD, and the required output voltage is set, as in the measurement of the sensitivity. Then, by upsetting the SSG in both directions relative to the tuning frequency of the receiver and increasing the input signal level, a normal output voltage is achieved at the RRD output. Based on the data obtained, a single-signal selectivity characteristic is constructed, according to which the passband of the radio control system, the squareness coefficient of the selectivity characteristic and the attenuation of the receiver along the adjacent receiving channel, the detuning value of which is determined in the TC, are determined. In this case, the attenuation of the adjacent reception channel is understood as a value that shows how many times the sensitivity of the receiver deteriorates at a given detuning corresponding to the adjacent reception channel, in relation to the sensitivity of the receiver. In the process of measurements, it should be borne in mind that sometimes with a wide preselector bandwidth, an increase in the input signal level can overload some of the receiver stages. In these cases, the level of the signal initially applied to the receiver input should be chosen less than the receiver sensitivity and should be chosen so that receiver overload is not observed. However, in this case, inherent noise can make measurements difficult. In such cases, an additional narrow-band filter is used at the output of the receiver to help suppress the noise floor and perform measurements [8].

The block diagram for taking the characteristics of the single-signal selectivity of the FSK receiver is somewhat different from the one discussed above, which is explained by the presence of amplitude limiters in the FSK receivers. Here, the measuring device at the output of the channel of the last IF amplifier and after selecting the level of the input unmodulated signal based on the absence of overload of the cascades, the selectivity characteristic is removed similarly to ASK receivers.

Measurement of one-signal selectivity of television receivers is carried out only for the image channel. In this case, the voltage of the carrier frequency of the image of the corresponding channel is applied to the input of the receiver, modulated by a voltage of frequency 400 or 1000 Hz with a modulation depth of 50% (or a signal with the parameters specified in the specifications). The controls of the television receiver and the SSG adjust and determine the gain-limited sensitivity. Then the same value is sequentially determined at the values of the selected detuning's and the results are used to build a characteristic of one-signal selectivity. In this case, as in the case of measuring the sensitivity, it is necessary to exclude the influence of the radio control system.

When determining the single-signal selectivity of microwave receivers, standard signal generators do not allow, as a rule, to remove the selectivity characteristic when the signal source is switched on directly to the receiver input, since the frequency readout accuracy of the generator detuning in the microwave range is low. Therefore, in such receivers, the one-signal selectivity characteristic is usually taken when the signal source is switched on directly to the input of the last intermediate frequency path (with a single frequency conversion, the signal source is connected to the input of the intermediate frequency path). In this case, the influence on the selectivity of the previous, to the point of switching on the SSG, receiver stages is excluded, however, the measurement error will be small, since these stages have little effect on the resulting characteristic of the receiver. This is due to the fact that the characteristic of single-signal selectivity of microwave receivers is mainly determined by the selectivity of the most narrow-band last channel of the IF amplifier [9].

When measuring the attenuation of signals on the side receiving channels, first measure the sensitivity of the receiver on the main receiving channel. Then, with a constant position of the receiver controls, the sensitivity of the radio receiver is measured at the frequencies of the side channels. Based on the data obtained, the amount of attenuation of the corresponding side receiving channel (SRC) is calculated as

$$\sigma_{SRC} = \frac{U_{SRC}}{E_{MRC}} \text{ или } \sigma_{ПК} = \frac{P_{Д}}{P_{A0}} \tag{7.2}$$

where U_{SRC}, (P_{SRC}) is the receiver sensitivity for this particular side channel;

E_{MRC}, (P_{MRC}) is the receiver sensitivity for the main receiving channel (MRC).

Note that in band receivers, the measurement of the attenuation of the side receiving channels should be performed at the frequency of each subband at which the attenuation is minimal.

7.3.2 Two-Signal Selectivity

The measurement of two-signal selectivity is performed according to the scheme shown in Fig. 7.2 and differs from the one-signal selectivity measurement scheme by the presence of the second SSG-2.

When the characteristics of the frequency selectivity of the receiver for blocking are taken, the SSG-I of the useful signal with the modulation parameters required by

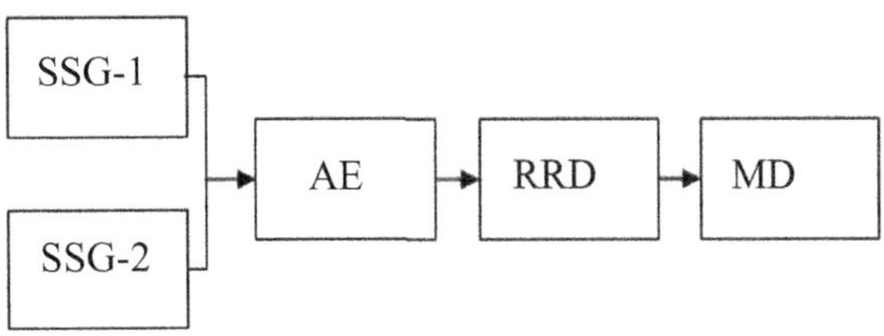

Fig. 7.2 The measurement of two-signal selectivity

the TC is switched on and the operations corresponding to the measurement of the receiver's sensitivity are performed. Then the SSG-2 is turned on with the modulation turned off and it is tuned to a frequency that differs by a certain amount from the tuning frequency of the receiver (for example, to the frequency of the adjacent receive channel). The signal level from the SSG-2 is selected so that the value of the output voltage of the receiver changes (decreases) by the value specified in the specification (usually 3 or 6 dB) compared to the nominal value. Then the detuning of the second generator is increased and the measurements are repeated. The obtained data is used to construct the characteristic of two-signal blocking selectivity.

When the characteristics of the frequency selectivity of the receiver for cross-distortion are taken, the SSG-I of the useful signal with the modulation parameters required by the TC is switched on and the operations corresponding to the measurement of the receiver sensitivity are performed. After that, turn off the modulation of the SSG-I signal and turn on the SSG-2, which simulates the interference signal with the modulation parameters specified in the TC. By changing the tuning frequency of the SSG-2, the level of interference is determined at which the value of the output voltage of the radio control system will change (decrease) for the value specified in the specification (for example, 3 dB) compared to the nominal value. Based on the obtained data, the ratio of the interference level to the receiver sensitivity is calculated at a given detuning and the characteristic of the receiver cross-distortion selectivity is constructed. When measuring the frequency selectivity of a radio receiver by intermodulation, the receiver is first tuned to the frequency of one of the SSG and the operations corresponding to the measurement of the sensitivity are performed, setting the required signal level at the output of the receiver. Then, with the help of both SSG, interference is simulated, the frequencies of which, after conversion, form a frequency equal to the tuning frequency f_{tf} of the receiver or the intermediate frequency f_{if}. To do this, use a combination of the frequencies of the SSG-1 and SSG-2 generators:

$$f_1 \pm f_2 = f_{if};\; f_1 \pm f_2 = f_{tf};\; 2f_1 - f_2 = f_{tf};\; 2f_2 - f_1 = f_{tf}.$$

Next, the generators are tuned to the frequencies of possible combinations and the same output noise levels are set. These noise levels at the input of the receiver are increased until the output level reaches the previously set nominal value. Then build the characteristic of the selectivity of the receiver for intermodulation, which is the dependence of the level of interference at the input of the receiver, creating intermodulation in it, on the frequency of one of the SSG at a given intermodulation coefficient.

7.4 Measuring the Distortion of Signals

When determining the degree of signal distortion in receivers, the harmonic coefficient is most often measured, the modulation characteristic, the fidelity characteristic and the transient amplitude characteristic are removed.

The block diagram for measuring the harmonic distortion coincides with the diagram shown in Fig. 7.1. A signal with modulation parameters specified in the TC is fed to the input of the RRD and all operations corresponding to the sensitivity measurement are performed, and then the harmonic coefficient is determined with a nonlinear distortion meter. Sometimes, to assess the degree of nonlinear distortion, the dependence of the harmonic coefficient on the modulation coefficient and the dependence of the harmonic coefficient on the amplitude of the input signal are additionally taken at a constant value of the modulation depth [10]. The determination of the first 'dependence is carried out at a constant amplitude of the input signal corresponding to the sensitivity. The second dependence is usually removed at two values of the modulation coefficient: 0.3 and 0.9.

The modulation characteristic, which is the dependence of the output voltage of the RR on the modulation index, is taken for the receivers of frequency-modulated signals at a constant input voltage level equal to the sensitivity of the RR and the signal modulation parameters specified in the specifications.

When removing the fidelity characteristic, which is the dependence of the output voltage of the ASK receiver on the modulation frequency of the input signal at a constant signal level and modulation index, perform operations similar to measuring the sensitivity, and then change the modulation frequency within the limits specified in the specifications, and note the value of the output voltage ... Based on the data obtained, the fidelity characteristic is plotted and the signal attenuation coefficient at the upper and lower modulating frequencies is determined.

To determine the degree of distortion of impulse signals during their passage through the linear path of the RRD, the transient response of the amplitude of the path is removed. However, often instead of this characteristic, the degree of the resulting distortion of pulse signals at the output of the entire radio receiver is determined, i.e., taking into account the distortions introduced by the detector and the video amplifier, In this case, such indicators as the rise time and fall time of the pulses, the amplitude of surges, the value of the fall of the top of the pulse for its duration are measured. The block diagram for these measurements coincides with the diagram shown in Fig. 7.1. A pulse signal with the signal structure specified in the TC (duration and pulse repetition period, type of modulation) is supplied to the input of the RRD from the SSG, performs operations corresponding to the measurement of the sensitivity of the radio receiver, and determines the degree of signal distortion using the oscilloscope [9, 11].

7.5 Determination of Regulatory Characteristics

7.5.1 Automatic Gain Control

By a similar technique, the fidelity characteristic is also taken for the FM receiver. In this case, the modulation frequency is changed, leaving constant the frequency deviation and the signal level at the input of the RRD.

Removing the amplitude characteristic is carried out according to the structural diagram of the measurement of the real sensitivity of the receiver. The receiver (if it is a range one) is tuned to the frequency at which its sensitivity is highest. Further, control measurements of the sensitivity are performed. After that, without changing the position of the receiver controls, remove its amplitude characteristic with AGC and without AGC. The signal level from the SSG is changed from values significantly less than the sensitivity to a value at which the increase in the output voltage becomes much less than the increase in the input signal.

Sometimes the limits of the change in the output signal when the amplitude characteristic is taken are stipulated in the TC. Based on the characteristics taken, the dynamic range of signals at the input and output of the RRD is determined [12].

7.5.2 Automatic Frequency Control

The main parameters of the AFC system can be determined by its dynamic characteristics, the structural diagram for removing which coincides with the diagram shown in Fig. 7.1, where a frequency meter serves as a measuring device.

The sequence of measurements for static AFC systems can be as follows. A signal corresponding to the real sensitivity of the receiver is fed from the SSG to the input of the receiver. Next, turn off the AFC system. By changing the tuning frequency of the SSG and monitoring the frequency meter, set the nominal value of the intermediate frequency. Then turn on the AFC system. By giving the input signal an increment in frequency in the direction of its increase, the corresponding increments of the intermediate frequency are determined. The signal frequency is increased until the system leaves the tracking mode [13]. Similar measurements are performed in the direction of negative frequency increments of the input signal. Then measurements are made, reducing the initial detuning of the SSG from the side of large initial detunings to the frequency at which the system enters the auto-tracking mode, providing the specified residual error. Based on the data obtained, a dynamic error curve is constructed and from it the auto-tuning coefficient is found, the residual error value, the retention and capture band. For astatic AFC systems, only the hold-and-hold band is determined.

7.6 Additional Types of Tests

Determination of the level of spurious radiation of local oscillators is carried out with a noise meter at fundamental and combination frequencies.

The method of testing the stability of the local oscillator frequency is the interference method, which consists in comparing the frequency of the local oscillator with the frequency of some reference oscillator of the corresponding stability. This test uses the radiation of the local oscillator acting on the interference wave meter. The generator of this wave meter is tuned by zero beats to the local oscillator frequencies by the absolute or relative value of the deviation from zero beats, and the stability of the heterodyne frequency is determined.

The characteristic of the noise is taken at the same tuning frequencies of the receiver, at which its sensitivity was measured. To characterize the noise, a number of values of an unmodulated high-frequency voltage with a frequency equal to the tuning frequency of the receiver are fed to the input of the receiver. For each value of the input voltage, the noise voltage at the output of the receiver is measured, and based on the obtained values, a noise characteristic is constructed.

Power supplies, structurally combined with the receiver, are tested together with the receiver.

To assess the relative changes in the voltage of rectifiers, both voltage instability meters and various special measurement schemes are used, in which compensation methods are used, which consist in comparing the measured voltage with a reference constant voltage. A feature of the measuring devices is that they allow you to measure the internal resistances of the sources and the relative change in the investigated voltage per unit of time.

The ripple voltage is measured using an oscilloscope, which is pre-calibrated by voltage from a reference source [14].

7.7 Test Automation

In connection with the emergence of a wide class of discrete and digital measuring and computer technology, the possibilities for automating the monitoring of the operability of receivers and measuring their characteristics have significantly expanded. In modern radio control systems, parameter control devices are either an integral part of them, or represent separate devices designed to control the parameters of receivers of a certain class and purpose.

Along with a significant reduction in the time required for control and measurement, the automation of the processes of measuring parameters leads to significant changes in the requirements for the design, structural and schematic diagrams of the radio control room in the sense of providing conditions for connecting measuring and controlling automata. The system of automatic control and measurement of parameters is a set of separate functional blocks and nodes that generate input signals, measure and process output signals, record measurement results and control the operation of the control system.

Automatic control devices can be designed both for control and measurement of one or several parameters, and for control of the entire set of parameters specified in the technical specifications. In this case, in any case, the measurement and control of the characteristics of the receiver are based on the methods discussed in the previous subsections.

The structural diagram and the set of functional blocks of the control machine depend on the purpose of using the machine and the set of monitored and measured parameters [12, 15]. In Fig. 7.3, for example, a block diagram of one of the possible variants of a control device is shown, which allows automating the processes of measuring the sensitivity, degree of nonlinear distortion, attenuation of additional reception channels, taking the amplitude-frequency and amplitude characteristics of the receiver, etc. The machine can measure and record all monitored parameters or the same measurement of all parameters and registration of only those that do not fit into the permissible limits.

The considered control machine consists of separate blocks. Voltage generators of low frequency (GLF) and high frequency (GHF) are connected to the input of the receiver (PR) through a switch K_1 and attenuators ATT-I and ATT-2 of low and high frequency, respectively. The output voltage from the receiver goes to an electronic switch K_2, which, upon a command from the control device (CD), connects the receiver output to a high-pass filter (HPF) or to a switch K_3. Voltage with K_3, depending on the measured parameter, is supplied to a linear amplitude detector (LAD) or a detector with a square-law characteristic (SLC). The outputs of the detectors are connected to a voltage meter (VM). The measurement results are transmitted

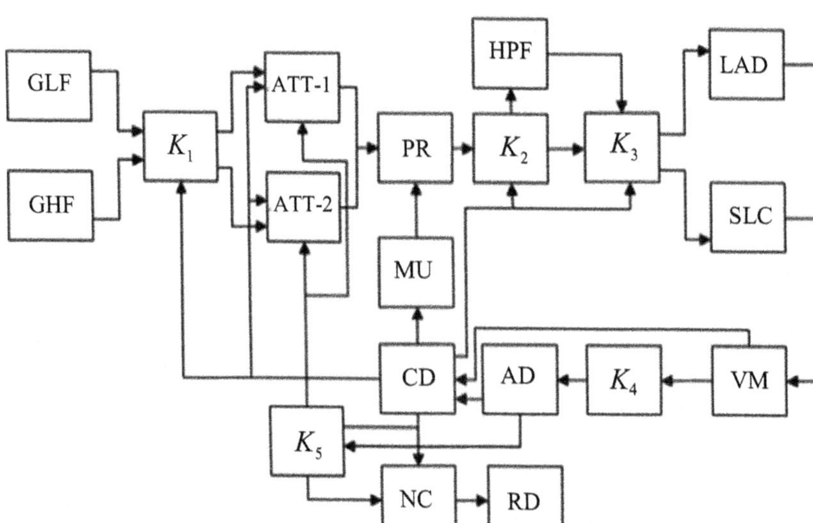

Fig. 7.3 A block diagram of one of the possible variants of a control device

either through the control unit to the recording device (RD), or through the switch K_4 to the arithmetic device (AD) for data processing, and then through the switch K_5 to the recording device. A number comparator (NC) is included between the K_1 and input device, which serves to compare the measurement results of this parameter with the value specified in the specification and the permissible deviations from it. The mechanics unit (MU) performs mechanical operations according to commands from the control unit.

Let us consider, using the example of a control device (Fig. 7.3), the sequence of determining some characteristics of the receiver.

When measuring the sensitivity, a standard signal is supplied to the receiver input from the GHF through the switch K_1 controlled by the CD and the ATT-2 attenuator. The signal level required by TC is set by ATT-2 on command from the control CD unit. Switches K_2 and K_3 supply the output voltage of the receiver to the LAD. The rectified voltage from the LAD is fed to the VM, which measures the output voltage in the absence and presence of a useful signal at the receiver input. The measurement results are sent to the arithmetic device and, after comparison through the AD, to the recording device.

The harmonic distortion measurement is usually performed at one modulating frequency for a given value of the receiver output voltage. A signal with a given modulation frequency is fed to the input of the receiver from the GHF and the ATT-2 attenuator sets the required output voltage level of the PR, which is fed through to the filter HPF. The switch K_3 connects the SLC detector with a square-law characteristic to the output HPF. The output voltage from the SLC is fed to the VM. The measurement results are sent to the AD. Then, the full output voltage from the PR is applied to the SLC through K_2 and K_3 is supplied. The result of the second measurement is also fed to the AD. Comparison of both measurements allows you to determine the harmonic distortion.

As can be seen from the examples discussed above, in such an automatic receiver parameter meter, the general principles of measuring the main characteristics of receivers are implemented. However, the measurements are automated [10, 13, 15].

Most frequencies in practice use a sequential control method. In this case, the principle of operation of the control device can be different. In particular, the principle of constructing a control machine is possible, when the control or measurement of the next parameter occurs only if the value of the measured parameter is within the limits defined in the technical specifications, or the control of the next parameter begins regardless of the results of measurement of the previous parameters. Note, however, that a sequential measurement has the significant disadvantage that the measurements take a long time. However, simultaneous, parallel control of several parameters is possible only if the monitored paths can work autonomously and different functional blocks of the control machine are used for measurements.

References

1. Belousov AP (1959) Calculation of the noise figure of radio receivers. M.: State publishing house of defense. prom-sti, p 135.
2. Goldenberg LM, Matyushkin BD, Polyak MN (1985) Digital signal processing: A handbook. M.: Radio and communication, p 312.
3. Gutkin LS, Chentsova OS (1958) Transient processes in the system: high-frequency amplifier—detector. Radio engineering, no. 11, pp 18–23.
4. Ditkin VA, Prudnikov AP (1965) A guide to operational calculus. M.: Higher. shk., p 466.
5. Evtyanov SI (1948) Transient processes in receiving-amplifying circuits. M.: Radio and communication, p 210.
6. McElroy JH (1977) Communication systems for near space using CO_2 lasers. TIIER, no. 2, pp 54–89.
7. Malashin MS, Kaminsky RP, Borisov YB (1983) Basics of designing laser locating systems: Textbook manual for universities. M.: Higher. shk., p 207.
8. Martynov VA, Selikhov YI (1980) In: Zavarin GD (ed) Pan-frame receivers and spectrum analyzers. M.: Sov. radio, p 352.
9. Pervachev SV (1982) Radioautomatics: Textbook for universities. M.: Radio and communication, p 296.
10. Pestryakov VB, Kuzenkov VD (1985) Radio engineering systems. M.: Radio and communication, p 376.
11. Poberezhsky ES (1987) Digital radio receivers. M.: Radio and communication, p 184.
12. Bankov VN, Barulin LG, Zhodzishsky MI, et al. (1984) In: Barulina LG (ed). Radio receivers. M.: Radio and communication, p 272.
13. Korostelev AA, Klyuev NF, Miller YA, et al. (1978) In: Dulevich VE (ed) Theoretical foundations of radar: Textbook for universities. M.: Soviet radio, p 608.
14. Zhodzishsky MI, Sila-Novitsky SY, Prasolov VA, et al. (1980) In: Zhodzishsky MI (ed) Digital phase synchronization systems. M.: Soviet radio, p 208.
15. Chistyakov NI (1938) To the calculation of one discriminator circuit. News of the electrical industry of a weak current, no. 2.

The manufacturer's authorised representative in the EU is Springer Nature Customer Service Centre GmbH, Europaplatz 3, 69115 Heidelberg, Germany. If you have any concerns regarding our products, please contact ProductSafety@springernature.com

Printed and bound by CPI Group (UK) Ltd, Croydon, CR0 4YY

30/09/2026

02225308-0011